RevisionGuide

GCSEBiology

**Jackie Clegg and
Mike Smith**

Series editor: Jayne de Courcy

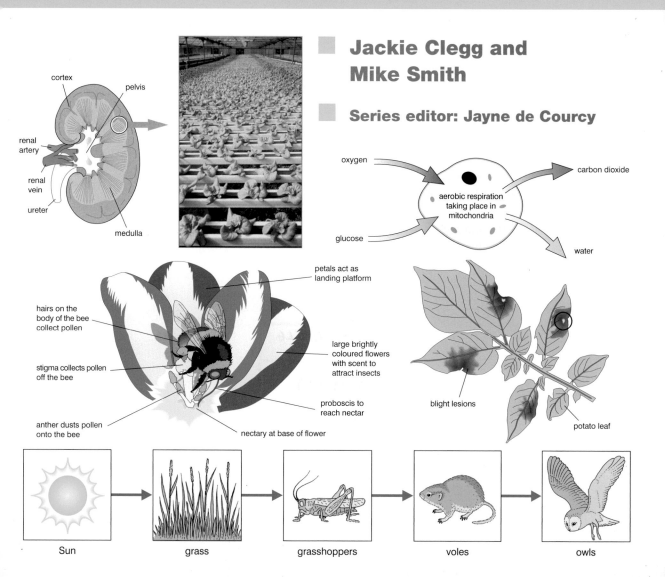

CONTENTS AND REVISION PLANNER

There is a slight variation in the content of the Biology taught for the different exam boards. The table shows you which pages you should pay particular attention to according to your exam board.

AQA	Edexcel	OCR
25	52	14
26	53	26
27		30
28		44
29		76
48		77
		97

ABOUT THIS BOOK

Exams are about much more than just repeating memorised facts, so we have planned this book to make your revision as active and effective as possible.

How?

- by breaking down the content into manageable chunks (Revision Sessions)

- by testing your understanding at every step of the way (Check Yourself Questions)

- by providing extra information to help you aim for the very top grade (A* Extras)

- by highlighting Ideas and Evidence topics (Ideas and Evidence)

- by suggesting the most likely exam questions for each topic (Question Spotters)

- by giving you invaluable examiners' guidance about exam technique (Exam Practice)

REVISION SESSION I

Revision Sessions

- Each unit is divided into a number of **short revision sessions**. You should be able to read through each of these in no more than 30 minutes. That is the maximum amount of time that you should spend on revising without taking a short break.

- Ask your teacher for a copy of your own exam board's **GCSE Biology specification**. Match the topics in it with this book's Contents list, ticking off in the box headed '**On specification**' each of the revision sessions that contains the topics you need to cover. It will probably be most of them.

CHECK YOURSELF QUESTIONS

- At the end of each revision session there are some **Check Yourself Questions**. By trying these questions, you will immediately find out whether you have understood and remembered what you have read in the revision session. **Answers** are at the back of the book, along with **extra hints and guidance**.

- If you manage to give correct answers to all the Check Yourself questions for a session, then you can confidently tick off this topic in the box headed '**Revised & understood**'. If not, you will need to tick the '**Revise again**' box to remind yourself to return to this topic later in your revision programme.

A* EXTRA

These boxes occur in each revision session. They contain some **extra information** which you need to learn if you are aiming to achieve the very top grade. If you have the chance to use these additional facts in your exam, it could make the difference between a good answer and a very good answer.

IDEAS AND EVIDENCE

Biology GCSE specifications have **particular topics highlighted** as 'ideas and evidence'. Every Foundation and Higher Tier paper must have at least one question on one of these topics.

The boxes in the book highlight the sort of questions you may meet in relation to themes such as the development of scientific ideas, the applications of biology, environmental issues and economic factors.

QUESTION SPOTTER

It's obviously important to revise the facts, but it's also helpful to know **how you might need to use this information** in your exam.

The authors, who have been involved with examining for many years, know the sorts of questions that are most likely to be asked on each topic. They have put together these Question Spotter boxes so that they can help you to **focus your revision**.

- This unit gives you **invaluable guidance on how to answer exam questions well.**

- It contains some sample students' answers to typical exam questions, followed by examiners' comments on them, showing where the students gained and lost marks. Reading through these will help you get a very clear idea of what you need to do in order to score **full marks** when answering questions in your GCSE Biology exam.

- There are also some **typical exam questions** for you to try answering. Model answers are given at the back of the book for you to check your own answers against. There are also examiners' comments, highlighting **how to achieve full marks**.

About your GCSE Biology course

Does this book match my course?

This book has been written to support all the single-award Biology GCSE specifications produced by the four examining groups in England and Wales. These specifications are based on the National Curriculum Key Stage 4 Programmes of Study. Ask your teacher for a copy of the specification you are following so that you can use it as a check-list.

Foundation and Higher Tier papers

In your GCSE Biology exam you will be entered for either the Foundation Tier exam papers or the Higher Tier exam papers. The Foundation exams allow you to obtain grades from G to C. The Higher exams allow you to obtain grades from D to A*.

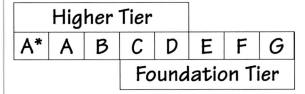

	Higher Tier						
A*	A	B	C	D	E	F	G
			Foundation Tier				

What will my exam questions be like?

The exam questions will be of a type known as structured questions. Usually these are based on a particular topic and will include related questions. Some of these questions will require short answers involving a single word, phrase or sentence. Other questions will require a longer answer involving extended prose. You will have plenty of practice at both types of questions as you work through this book.

Short answer questions

These are used to test a wide range of knowledge and understanding quite quickly. They are often worth one mark each.

Extended prose questions

These are used to test how well you can link different ideas together. Usually they ask you to explain ideas in some detail. It is important to use the correct scientific terms and to write clearly. They may be worth four or five marks, and sometimes more.

Ideas and Evidence questions

How scientific ideas grow, and how they are communicated, will form part of your exam. Often this means that quite a lot of information is given in the question. *Don't panic over such questions:* you may not be familiar with the particular example in the question, but the point is that a number of themes apply to scientific development in general. Use the Ideas and Evidence boxes in this book to see examples of these themes.

Quality of written communication

There will be marks for this in your exam. Remember to use capital letters at the start of sentences and full stops at the end. Check your spelling carefully, and try to use the correct technical words where you can.

How should I answer exam questions?

- Look at the *number of marks*. The marks should tell you how long to spend on a question. A rough guide is a minute for every mark. The number of marks will indicate how many different points are required in the answer.
- Look at the *space allocated for the answer*. If only one line is given, then only a short answer is required, e.g. a single word, a short phrase or a short sentence. Some questions will require more extended writing and for these four or more lines will be allocated.
- *Read the question carefully.* Students frequently answer the question they would like to answer, rather than the one that has actually been set! Circle or underline the key words. Make sure you know what you are being asked to do. Are you choosing from a list? Are you completing a table? Have you been given the formula you will need to use in a calculation? Are you describing or explaining?

UNIT 1: LIFE PROCESSES AND CELLS

Characteristics of living things

☐ Cells

- Almost all living things (**organisms**) are made up of building blocks called **cells**. Most cells are so small that they can only be seen with a microscope.

- Cells help organisms carry out some of the processes that are vital for life. There are seven processes needed for life. The easy way to remember all seven processes is to take the first letter from each process. This spells **Mrs Gren**:
  - **m**ovement
  - **r**espiration
  - **s**ensitivity
  - **g**rowth
  - **r**eproduction
  - **e**xcretion
  - **n**utrition.

- Plant cells have features that are not found in animal cells. The diagrams below show a typical animal cell and typical plant cells.

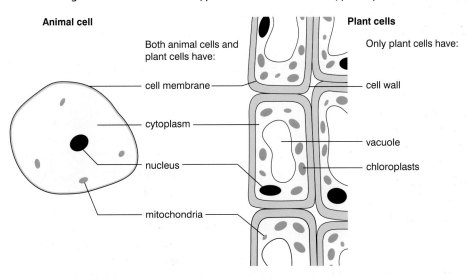

Animal cell

Both animal cells and plant cells have:

cell membrane

cytoplasm

nucleus

mitochondria

Plant cells

Only plant cells have:

cell wall

vacuole

chloroplasts

> ⚡ **A* EXTRA**
>
> ▸ The cell membrane contains tiny holes. This allows it to be **partially permeable**.

☐ What do the different parts of a cell do?

- The **cell membrane** holds the cell together and controls substances entering and leaving the cell.

- The **cytoplasm** is more complicated than it looks. It is where many different chemical processes happen. The cytoplasm contains **enzymes** that control these chemical processes.

- The **nucleus** contains **chromosomes** and **genes**. These control how a cell grows and works. Genes also control the features that can be passed on to offspring.

■ The **mitochondria** (singular: mitochondrion) are the parts of the cell where energy is released from food in the process known as **respiration**.

■ Plant cells have a **cell wall** as well as a cell membrane. The cell wall is made of **cellulose** and gives the cell shape and support.

■ A plant cell's **vacuole** contains a liquid called cell sap, which is water with various substances dissolved in it for storage. In a healthy plant the vacuole is large and helps support the cell.

■ Plant cells also contain **chloroplasts**. These contain the green pigment **chlorophyll**, which absorbs the light energy that plants need to make food in the process known as **photosynthesis**.

Similarities	Differences	
Both animal and plant cells contain	**Animal cells**	**Plant cells**
Cell membrane	Have many irregular shapes	Usually have a regular shape
Cytoplasm	Have no cell wall	Have a cell wall
Nucleus	Have no large vacuole (although	Usually have a large vacuole
Mitochondria	there may be small temporary vacuoles)	
	Have no chloroplasts	Green parts of a plant contain chloroplasts

⬚ Organisation of cells

■ The life processes are carried out by the co-ordinated action of cells. Cells of the same type, carrying out the same job, form **tissue**. Different tissues may make up an **organ**. Organs and tissues work together to make up an **organ system**. For example:

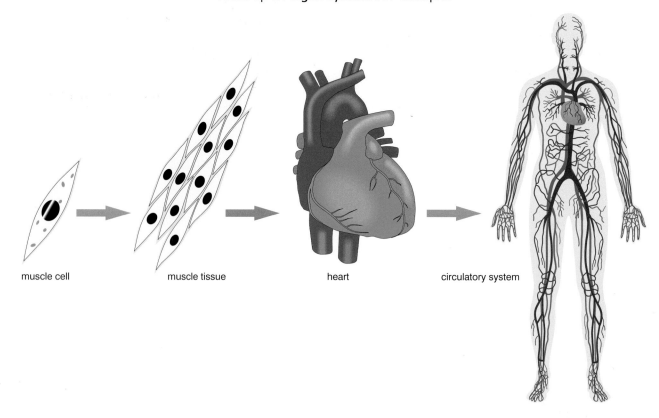

muscle cell   muscle tissue   heart   circulatory system

■ Different types of cells carry out different jobs. Cells have special features that allow them to carry out their job. This is called **specialisation**. Good examples of specialised cells are sperm cells, nerve cells, blood cells and root hair cells.

IDEAS AND EVIDENCE

▸ In 1665 Robert Hooke invented the first useful microscope. He used it to look at thin slices of cork. He saw little box-like shapes that he called cells because he thought they looked like the tiny rooms or cells in which monks lived.

▸ The invention of the microscope allowed scientists to develop a greater understanding of living organisms.

CHECK YOURSELF QUESTIONS

Q1 Look at the diagram of an onion cell.

a How is the onion cell different from a typical plant cell?
b Explain the reason for this difference.
c How is the onion cell different from an animal cell?

Q2 Name which part of a cell does the following.
a Releases energy from food.

b Allows oxygen to enter.
c Contains the genes.
d Contains cell sap.
e Stops plant cells swelling if they take in a lot of water.

Q3 For each of the following say whether it is a cell, a tissue, an organ, a system or an organism:
a oak tree
b human egg
c brain, spinal cord and nerves
d leaf
e stomach lining
f kidney.

Answers are on page 136.

Transport into and out of cells

How do substances enter and leave cells?

■ The cell membrane is **partially permeable** and this allows substances to move into and out of the cell.

■ There are three main ways substances enter and leave cells:

  1 diffusion

  2 osmosis

  3 active transport.

Diffusion

■ Substances like water, oxygen, carbon dioxide and food are made of particles (**molecules**).

■ In liquids and gases the particles are constantly moving around. This means that they will tend to spread themselves out evenly. For example, if you dissolve sugar in a cup of tea, even if you do not stir it, the sugar will eventually spread throughout the tea because the sugar molecules are constantly moving around, colliding with and bouncing off other particles. This is an example of **diffusion**.

> Diffusion is net movement from a region of high concentration to a region of low concentration.

 water molecule

 sugar molecule

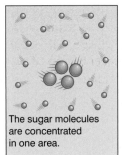
The sugar molecules are concentrated in one area.

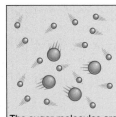
The sugar molecules are spreading out because they are constantly moving and colliding.

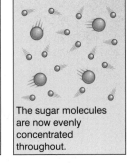
The sugar molecules are now evenly concentrated throughout.

SUMMARY OF DIFFUSION

■ Diffusion occurs when there is a **difference** in concentration. The greater the difference in concentration the faster the rate of diffusion.

■ Particles diffuse from regions of high concentration to regions of low concentration. They move down the **concentration gradient**.

■ Diffusion stops when the particles are evenly concentrated. But this does not mean that the particles themselves stop moving.

■ Diffusion happens because particles are constantly and randomly moving. It does not need an input of energy from a plant or animal.

■ Larger particles (like sugar molecules) diffuse through membranes more slowly than smaller particles (like oxygen).

DIFFUSION IN CELLS

- Substances can enter and leave cells by diffusion. If there is a higher concentration on one side of the membrane than the other and the substance can move through the membrane, then it will.

- For example, red blood cells travel to the lungs to collect oxygen. There is a **low** oxygen concentration in the red blood cells (because they have given up their oxygen to other parts of the body) and a **high** oxygen concentration in the alveoli of the lungs. Therefore oxygen diffuses into the red blood cells.

- Other examples of diffusion include:

  - carbon dioxide entering leaf cells

  - digested food substances in the small intestine entering the blood

  - in a kidney dialysis machine, the movement of urea from the dialysis tube into the dialysis fluid.

Osmosis

- **Osmosis** is a special example of diffusion where **only water** molecules move into or out of cells. It occurs because cell membranes are **partially permeable**: they allow some substances (such as water) to move through them but not others.

- Water molecules will diffuse from a place where there is a high concentration of water molecules (such as a dilute sugar solution) to where there is a low concentration of water molecules (such as a concentrated sugar solution).

- Many people confuse the concentration of the solution with the concentration of the water. Remember, it is the **water molecules that are moving**, so you must think of the concentration of **water molecules in the solution** instead of the concentration of substance dissolved in it.

  - A low concentration of dissolved substances means a high concentration of water molecules.

  - A high concentration of dissolved substances means a low concentration of water molecules. So the water molecules are still moving from a **high concentration (of water molecules)** to a **low concentration (of water molecules)**, even though this is often described as water moving from a low-concentration solution to a high-concentration solution.

- One important example of osmosis is water entering the roots of plants.

- Water molecules enter a plant cell until the cytoplasm pushes against the cell wall. The cell is then said to be **turgid**. Turgid cells are important for supporting the plant.

- If a plant is losing water faster than it can absorb water from the soil, the plant cells will become **flaccid**. This is what is happening when a plant wilts.

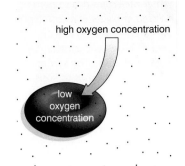

A high concentration gradient leads to a faster rate of diffusion. Red blood cells low in oxygen will absorb oxygen from oxygen-rich surroundings.

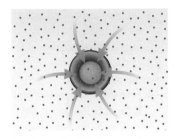

A red blood cell in pure water.

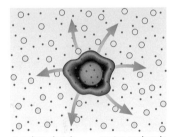

○ salt molecule   · water molecule

A red blood cell in a concentrated salt solution.

QUESTION SPOTTER

- You will be expected to apply your knowledge of osmosis to answer a question. For example, you may have to explain why plants wilt when grown in dry conditions.
- Try to include the key words **turgid** and **flaccid** in your answer.

Water molecules diffuse from an area of higher concentration (of water molecules) into an area of lower concentration (of water molecules). This sort of diffusion is known as osmosis.

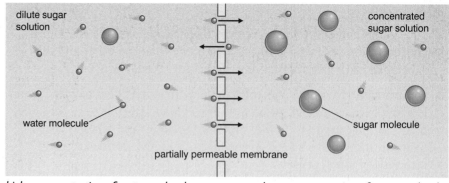

dilute sugar solution

concentrated sugar solution

water molecule

sugar molecule

partially permeable membrane

higher concentration of water molecules   *lower concentration of water molecules*

■ If animal cells such as blood cells are placed in different strength solutions they will shrink or swell up (and even burst) depending on whether they gain or lose water molecules via osmosis.

⬚ Active transport

■ Sometimes cells need to absorb particles **against a concentration gradient**: from a region of low concentration into a region of high concentration.

■ For example, root hair cells may take in nitrate ions from the soil even though the concentration of these ions is higher in the plant than in the soil. The way that the nitrate ions are absorbed is called **active transport**. Another example of active transport is dissolved ions being actively absorbed back into the blood from kidney tubules.

■ Active transport occurs when special **carrier proteins** on the surface of a cell pick up particles from one side of the membrane and transport them to the other side. You can see this happening in the diagram below.

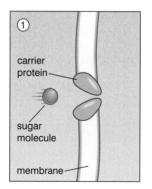

① carrier protein

sugar molecule

membrane

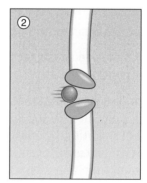

②

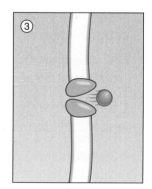

③

■ Active transport uses energy that the cells release during respiration.

CHECK YOURSELF QUESTIONS

Q1 An old-fashioned way of killing slugs in the garden is to sprinkle salt on them. This kills slugs by drying them out. Explain why this would dry them out.

Q2 Which of the following are examples of diffusion, osmosis or neither?
  a Carbon dioxide entering a leaf when it is photosynthesising.
  b Food entering your stomach when you swallow.
  c Tears leaving your tear ducts when you cry.
  d A dried out piece of celery swelling up when placed in a bowl of water.

Q3 Why would you expect plant root hair cells to contain more mitochondria than other plant cells?

Q4 Describe the difference between diffusion and active transport. Include one example of each in your answer.

Answers are on page 136.

UNIT 2: HUMAN BODY SYSTEMS

Cells and respiration

What is respiration?

■ The energy that our bodies need to keep us alive is released from our food. Releasing the energy is called **respiration** and happens in every cell of our body. The food is usually **glucose** (sugar) but other kinds of food can be used if there is not enough glucose available.

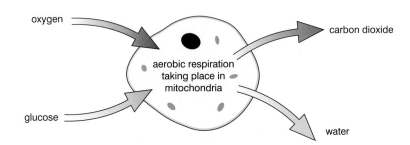

oxygen

carbon dioxide

aerobic respiration taking place in mitochondria

glucose

water

■ Different foods contain different amounts of **energy**. Carbohydrates (such as glucose) and proteins contain a similar amount of energy, but fat contains nearly twice the amount of energy.

■ The amount of energy in food is measured using a **calorimeter**. The amount of energy used to be measured in calories, but now it is measured in **joules**.

■ Respiration usually involves **oxygen**. This kind of respiration is called **aerobic respiration**. Water and carbon dioxide are produced as waste products. This is very similar to burning fuel. However, in our bodies enzymes control the rate at which energy is released.

■ Aerobic respiration can be summarised by a word equation:

glucose + oxygen → water + carbon dioxide + energy

■ It can also be written as a symbol equation:

$C_6H_{12}O_6 + 6O_2 \rightarrow 6H_2O + 6CO_2 + energy$

■ The important number to remember is **6**.

■ The energy released is used for all the other processes of life: movement, sensitivity, growth, reproduction, excretion and nutrition (feeding). Eventually most of the energy is lost as heat.

■ Aerobic respiration provides most of the energy we need. During exercise we need more energy so the rate of aerobic respiration increases:

• muscle cells need more glucose and oxygen

• more glucose is removed from the blood

- breathing becomes faster and deeper, to take in more oxygen

- the heart rate increases to deliver the oxygen and glucose to the muscle cells more quickly.

Anaerobic respiration

- There is a limit to how fast we can breathe and how fast the heart can beat. This means that the muscles might not get enough oxygen. In this case another kind of respiration is used that does not need oxygen. This is called **anaerobic respiration**.

- Anaerobic respiration can be summarised by a word equation:

glucose → lactic acid + energy

- Anaerobic respiration releases much less energy than aerobic respiration.

At the end of a race, athletes can be in a lot of pain because of the lactic acid released by anaerobic respiration in the muscle cells.

Differences between aerobic and anaerobic respiration	
Aerobic respiration	**Anaerobic respiration**
Uses oxygen	Does not use oxygen
Does not make lactic acid	Makes lactic acid
Makes carbon dioxide	Does not make carbon dioxide
Makes water	Does not make water
Releases a large amount of energy	Releases a small amount of energy

The oxygen debt

- The **lactic acid** that builds up during anaerobic respiration is poisonous. It causes muscle fatigue (tiredness) and makes muscles ache.

- Lactic acid has to be broken down, and oxygen is needed to do this. This is why you continue to breathe quickly even after you have finished exercising. You are taking in the extra oxygen you need to remove the lactic acid. This is sometimes called **repaying the oxygen debt**.

- Only when all the lactic acid has been broken down do your heart rate and breathing return to normal.

QUESTION SPOTTER

▸ You can expect to get at least one question on respiration in your exam.
▸ A key to answering most questions is remembering the equations. For example, you may be asked why your breathing rate increases during exercise. In your answer include the need for an increased oxygen supply so the muscles can release more energy, and the need to get rid of the extra carbon dioxide produced.

? CHECK YOURSELF QUESTIONS

Q1 Why do living things respire?

Q2 Where does respiration happen?

Q3 a Why is aerobic respiration better than anaerobic respiration?

b If aerobic respiration is better, why does anaerobic respiration sometimes happen?

Answers are on page 137.

Blood and the circulatory system

⌂ What are the functions of the blood?

■ **Blood** is the body's **transport system**, carrying materials from one part of the body to another. Some of the substances transported are shown in the table below.

Substance	Carried from	Carried to
Food (glucose, amino acids, fat)	Small intestine	All parts of the body
Water	Intestines	All parts of the body
Oxygen	Lungs	All parts of the body
Carbon dioxide	All parts of the body	Lungs
Urea (waste)	Liver	Kidneys
Hormones	Glands	All parts of the body (different hormones affect different parts)

■ The blood also plays a part in **fighting disease** and in **controlling body temperature**. (You will find out more about these processes in Unit 4.)

Parts of the blood	Job
Plasma (pale yellow liquid making up most of the blood)	Transports food, carbon dioxide, urea, hormones, antibodies and other substances all dissolved in water. Heat is also redistributed around the body
Red blood cells	Carry oxygen (and some carbon dioxide)
White blood cells	Defend body against disease (see Unit 4)
Platelets	Involved in blood clotting

red blood cells

platelets

plasma

white blood cells

Blood is mostly water, containing cells and many dissolved substances.

Red blood cells

■ Red blood cells are **specialised** to carry oxygen.

Feature of red blood cells	How it helps
'Biconcave' disc shape (flattened with a dimple in each side)	Large surface area for oxygen to enter and leave
No nucleus	More room to carry oxygen
Contains haemoglobin (red pigment)	Haemoglobin combines with oxygen to form oxyhaemoglobin The oxygen is released when the cells reach tissues that need it
Small	Can fit inside the smallest blood capillaries. Small cells can quickly 'fill up' with oxygen as it is not far for the oxygen to travel right to the centre
Flexible	Can squeeze into the smallest capillary
Large number	Can carry a lot of oxygen

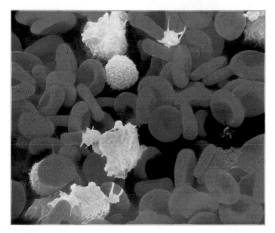

In this micrograph you can see the distinctive biconcave shape of red blood cells.

The circulatory system

■ Blood flows around the body through **arteries, veins** and **capillaries**. The **heart** pumps to keep the blood flowing.

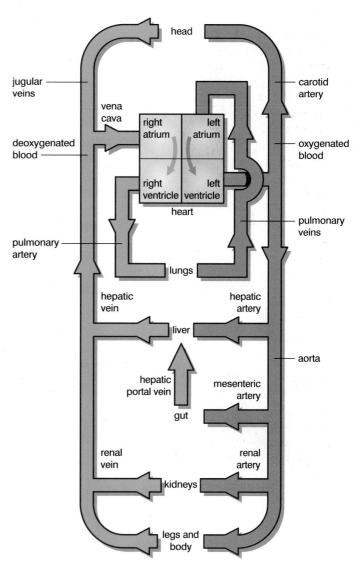

⊏⊐ The heart

■ The **heart** is a muscular bag that pumps blood by expanding in size, filling with blood, and then contracting, forcing the blood on its way.

■ The heart is two pumps in one. The right side pumps blood to the lungs to collect oxygen. The left side then pumps the **oxygenated blood** around the rest of the body. The **deoxygenated** (without oxygen) blood then returns to the right side to be sent to the lungs again.

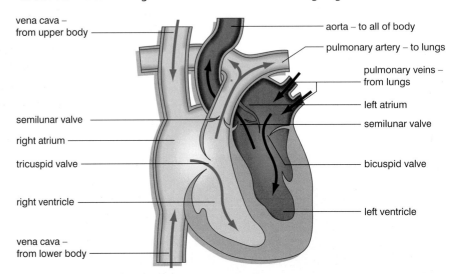

artery:

thick-walled carrying blood at high pressure

■ The heart contains several valves and four chambers, two called atria (singular: atrium) and two called ventricles. The **atria** have thin walls. They collect blood before it enters the ventricles. The **ventricles** have thick muscular walls that contract, forcing the blood out. The **valves** allow the blood to flow one way only, preventing it flowing back the way it came.

⊏⊐ Blood vessels

vein:

thin-walled carrying blood at low pressure

■ Blood leaves the heart through **arteries** and returns through **veins**. **Capillaries** connect the two. Remember, **a** for **a**rteries that travel **a**way from the heart. **V**eins carry blood into the heart and contain **v**alves.

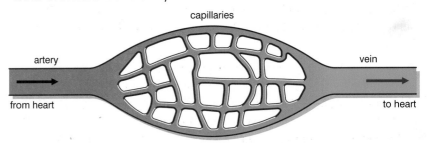

capillary:

very small; the walls may be just one cell thick

■ Arteries, veins and capillaries are adapted to carry out their different jobs.

■ Tissue fluid leaks from the capillaries. The tissue fluid bathes all the cells. Oxygen and other substances diffuse from the blood, through the tissue fluid to all of the cells. Waste substances such as carbon dioxide diffuse into the blood.

Blood vessel	Job	Adaptations	Explanation
Arteries	Carry blood away from heart	Thick muscular and elastic wall	Blood leaves the heart under high pressure. The thick wall is needed to withstand and maintain the pressure. The elastic wall gradually reduces the harsh surge of the pumped blood to a steadier flow
Veins	Carry blood back to the heart	Thinner walls than arteries	Blood is now at a lower pressure so there is no need to withstand it
		Large lumen (space in the middle)	Provides less resistance to blood flow
		Valves	Prevent back flow, which could happen because of the reduced pressure
Capillaries	Exchange substances with body tissues	Thin, permeable wall (may only be one cell thick)	Substances such as oxygen and food can enter and leave the blood through the capillary walls
		Small size	Can reach inside body tissues and between cells

normal blood flow

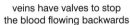

veins have valves to stop the blood flowing backwards

open

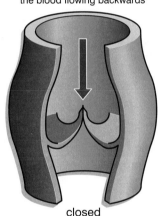

closed

⌗ Double circulation

- In humans (but not all animals) the blood travels through the heart twice on each complete journey around the body. This is a **double circulation**.

- By the time the blood has been pushed through a system of capillaries (in either the lungs or the rest of the body) it is at quite a low pressure. The pressure required to push the blood through the lungs and then the rest of the body in one go would be enormous and could damage the blood vessels. A double circulation system maintains the high blood pressure needed for efficient transport of materials around the body.

- The double circulation also allows for the fact that the pressure needed to push blood through the lungs (a relatively short round trip) is much smaller than the pressure needed to push blood around the rest of the body. This is why the left half of the heart is much more muscular than the right half.

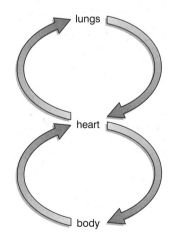

The heart pumps blood to the lungs and back, then to the body and back. This is a double circulation.

⌂ Blood pressure

- Your **blood pressure** is the measure of the pressure of your blood when your heart is contracting and relaxing.

- Increases in blood pressure can be caused by:

  - too much salt in the diet

  - increases in the levels of carbon dioxide in the blood because of exercise.

- The heart rate is adjusted to try and maintain a constant blood pressure. Nerves or hormones, such as adrenaline, bring about these adjustments. (You will find more about hormones in Unit 4.)

- Health risks are associated with blood pressure.

  - People with high blood pressure could suffer a stroke. This may be caused by blood vessels in the brain bursting.

  - People with low blood pressure may suffer with kidney problems, as the kidney loses its ability to filter blood.

? CHECK YOURSELF QUESTIONS

Q1 Small organisms, like *Amoeba*, a single-celled animal, do not need a transport system. Why do bigger organisms need them?

Q2 This equation shows the reversible reaction between oxygen and haemoglobin.

oxygen + haemoglobin ⇌ oxyhaemoglobin

  a Where in the body would oxyhaemoglobin form?

  b Where in the body would oxyhaemoglobin break down?

  c People who live at high altitudes, where there is less oxygen, have more red blood cells per litre of blood than people who live at lower altitudes. Suggest why.

Q3 **a** In the heart, the ventricles have thicker walls than the atria. Why is this?

  b Why does the left ventricle have a thicker wall than the right ventricle?

Q4 **a** List three ways that veins differ from arteries.

  b Substances such as oxygen and food enter and leave the blood through the capillary walls. Why do they not leave through the walls of arteries and veins?

Answers are on page 137.

Air and breathing

Breathing is *not* respiration

■ **Breathing** is the way that oxygen is taken into our bodies and carbon dioxide removed. Sometimes it is called **ventilation**.

■ Do not confuse breathing with respiration. Respiration is a chemical process that happens in every cell in the body. Unfortunately, the confusion is not helped when you realise that the parts of the body responsible for breathing are known as the **respiratory system**!

How we breathe

■ When we breathe, air is moved into and out of our lungs. This involves different parts of the respiratory system inside the **thorax** (chest cavity).

■ When we **breathe in**, air enters though the nose and mouth. In the nose the air is moistened and warmed.

■ The air travels down the **trachea** (windpipe) to the lungs. Tiny hairs called **cilia** help to remove dirt and microbes. (You will find out more about the cilia in Unit 4.)

■ The air enters the lungs through the **bronchi** (singular: bronchus), which branch and divide to form a network of **bronchioles**.

■ At the end of the bronchioles are air sacs called **alveoli** (singular: alveolus), which are covered in tiny blood capillaries. This is where oxygen enters the blood and carbon dioxide leaves the blood.

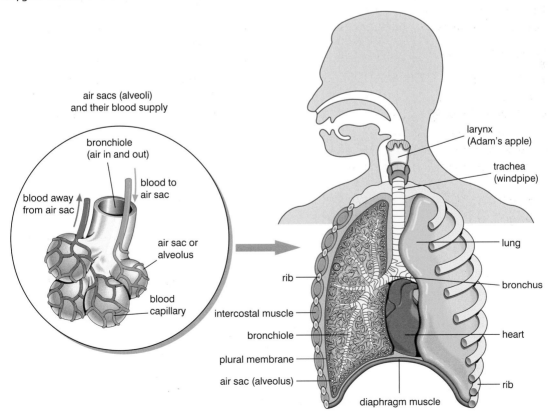

air sacs (alveoli) and their blood supply

bronchiole (air in and out)

blood to air sac

blood away from air sac

air sac or alveolus

blood capillary

larynx (Adam's apple)

trachea (windpipe)

lung

rib

bronchus

intercostal muscle

bronchiole

heart

plural membrane

air sac (alveolus)

rib

diaphragm muscle

► You need to understand how scientists are working to solve problems within our environment.

► Asthma is becoming a very common condition. During an asthma attack the walls of the bronchioles go into spasm and contract.

► Some of the things that are thought to cause an asthma attack include pollen, dust, animal hair and cold weather.

► Research is being carried out to try and understand why there has been a large increase in the number of people suffering from asthma.

Inhalation and exhalation

■ Breathing in is known as **inhalation** and breathing out as **exhalation** (sometimes they are called inspiration and expiration).

■ Both happen because of changes in the volume of the thorax. The change in volume causes pressure changes, which in turn cause air to enter or leave the lungs.

■ The changes in thorax volume are caused by the **diaphragm**, which is a domed sheet of muscle under the lungs, and the **intercostal muscles**, which connect the ribs. There are two sets of intercostals: the internal intercostal muscles and the external intercostal muscles.

INHALATION

■ Air is breathed into the lungs as follows.

1 The diaphragm **contracts** and **flattens** in shape.

2 The external intercostal muscles **contract**, making the ribs move upwards and outwards.

3 These changes cause the **volume** of the thorax to **increase**.

4 This causes the **air pressure** inside the thorax to **decrease**.

5 This causes air to **enter** the lungs.

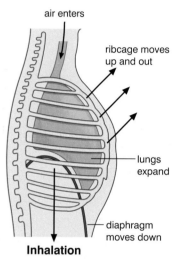

air enters
ribcage moves up and out
lungs expand
diaphragm moves down
Inhalation

■ Rings of **cartilage** in the trachea, bronchi and bronchioles keep the air passages open and prevent them from collapsing when the air pressure decreases.

EXHALATION

■ Air is breathed out from the lungs as follows.

1 The diaphragm **relaxes** and returns to its **domed** shape, pushed up by the liver and stomach. This means **it pushes up on the lungs**.

2 The external intercostal muscles **relax**, allowing the ribs to drop back down. This also presses on the lungs. If you are breathing hard the internal intercostal muscles also contract, helping the ribs to move down.

3 These changes cause the **volume** of the thorax to **decrease**.

4 This causes the **air pressure** inside the thorax to **increase**.

5 This causes air to **leave** the lungs.

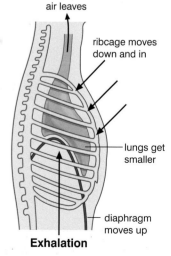

air leaves
ribcage moves down and in
lungs get smaller
diaphragm moves up
Exhalation

⚡ A* EXTRA

► During inhalation air enters because the air pressure inside the lungs is lower than the air pressure outside the body.
► During exhalation air leaves the lungs because the air pressure inside is higher than the air pressure outside the body.

☐ Composition of inhaled and exhaled air

- The air we breathe in and out contains many gases. **Oxygen** is taken into the blood from the air we breathe in. **Carbon dioxide** and **water vapour** are added to the air we breathe out. The other gases in the air we breathe in are breathed out almost unchanged, except for being warmer.

	In inhaled air	In exhaled air
Oxygen	21%	16%
Carbon dioxide	0.03%	4%
Nitrogen and other gases	79%	79%
Water	Variable	High
Temperature	Variable	High

☐ Alveoli

- The **alveoli** are where oxygen and carbon dioxide diffuse into and out of the blood. For this reason the alveoli are described as the site of **gaseous exchange** or the **respiratory surface**.

- The alveoli are **adapted** (have special features) to make them efficient at gaseous exchange. They have:

  - **thin, permeable walls** – to allow a short pathway for diffusion

  - a **moist lining** – in which oxygen dissolves first before it diffuses through

  - a **large surface area** – there are lots of alveoli, providing a very large surface area

  - a **good supply of oxygen** and **good blood supply** – which means that a concentration gradient is maintained so oxygen and carbon dioxide can rapidly diffuse across.

QUESTION SPOTTER

▸ You might be asked to explain how alveoli are adapted for efficient gas exchange.

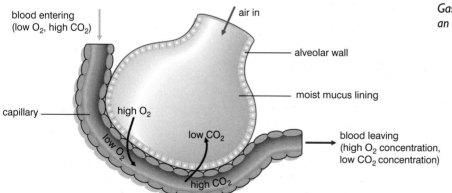

blood entering (low O$_2$, high CO$_2$)

air in

alveolar wall

moist mucus lining

capillary

high O$_2$

low O$_2$

low CO$_2$

high CO$_2$

blood leaving (high O$_2$ concentration, low CO$_2$ concentration)

Gaseous exchange in an air-filled alveolus.

- Smoking can damage the alveoli. (You will find out more about the links between smoking and respiratory diseases in Unit 5.)

Q1
a How many cells does oxygen pass through on its way from the alveoli to the red blood cells?

b Why is it important for there to be a large concentration gradient of oxygen between the inside of the alveoli and the blood?

Q2 When you breathe in, how do the positions of the diaphragm and ribcage change?

Q3 Why do we need rings of cartilage in the walls of the air passages?

Answers are on page 138.

Food and digestion

▢ What is a balanced diet?

- To keep us healthy we need a **balanced diet**. A balanced diet needs to include:

  - **proteins** – which are broken down to make amino acids, which are themselves used to form enzymes and other proteins needed by cells

  - **carbohydrates** – which are needed to release energy in our cells, to enable all the life processes to take place

  - **fat** – which is an important form of insulation to maintain body temperature, and is also used as a store of energy

  - **vitamins** and **minerals** – which are needed for the correct functioning of the body.

- Vitamins and minerals cannot be produced by the body and cooking food destroys some vitamins. This is why it is so important to eat raw fruit and vegetables.

Vitamins and minerals	Job	Good food source	Deficiency disease
Vitamin A	Helps cells to grow and keeps skin healthy, helps eyes to see in poor light	Liver, vegetables, butter, fish oil and milk	**Night blindness**
Vitamin B	For healthy skin and to keep the nervous system working properly	Meat, eggs, vegetables, fish and milk	**Beri-beri** (leg muscles are unable to grow properly)
Vitamin C	For healthy skin, teeth and gums, and keeps lining of blood vessels healthy	Citrus fruit and green vegetables	**Scurvy** (bleeding gums and wounds do not heal properly)
Vitamin D	For strong bones and teeth	Fish, eggs, liver, cheese and milk	**Rickets** (softening of the bones)
Calcium	Needed for strong teeth and bones, and involved in the clotting of blood	Milk and eggs	**Rickets** (softening of the bones)
Iron	Needed to make haemoglobin in red blood cells	Meat and spinach	**Anaemia** (reduction in number of red blood cells, person soon becomes tired and short of breath)

- Too much **saturated fat** (animal fat) causes **cholesterol** to be deposited inside blood vessels, making them narrower (as shown in the diagram on the right). The heart needs to work harder trying to push blood through these narrow vessels. This increases the risk of heart attack.

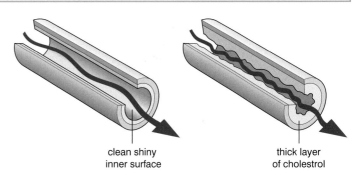

clean shiny inner surface

thick layer of cholestrol

- It is important to match your diet with:
  - how active you are
  - your age and gender
  - your body size.

IDEAS AND EVIDENCE

- It is useful to understand how early experiments were carried out.
- In 1774 James Lind was a ship's doctor on board a ship named the *Salisbury*. He carried out tests on sailors who were suffering with scurvy. Lind discovered that if sailors included oranges in their diet they quickly recovered.

	Energy used in a day (kJ)	
	Male	**Female**
6-year-old child	7500	7500
12–15-year-old teenager	12 500	9700
Adult manual worker	15 000	12 500
Adult office worker	11 000	9800
Pregnant woman		10 000

Why do we need to digest food?

- If the food we eat is to be of any use to us it must enter our blood so that it can **travel to every part of the body**.
- Many of the foods we eat are made up of **large, insoluble molecules** that would not easily enter the blood. This means they have to be **broken down into small, soluble molecules** that can easily enter and be carried dissolved in the blood. Breaking down the molecules is called **digestion**.
- There are two stages of digestion.
  1. **Physical digestion** occurs mainly in the mouth, where food is broken down into smaller pieces by the teeth and tongue. (You will find out more about teeth in Unit 3.)
  2. **Chemical digestion** is the breakdown of large food molecules into smaller ones.
- Some molecules, such as glucose, vitamins, minerals and water, are already small enough to pass through the gut wall and do not need to be digested.

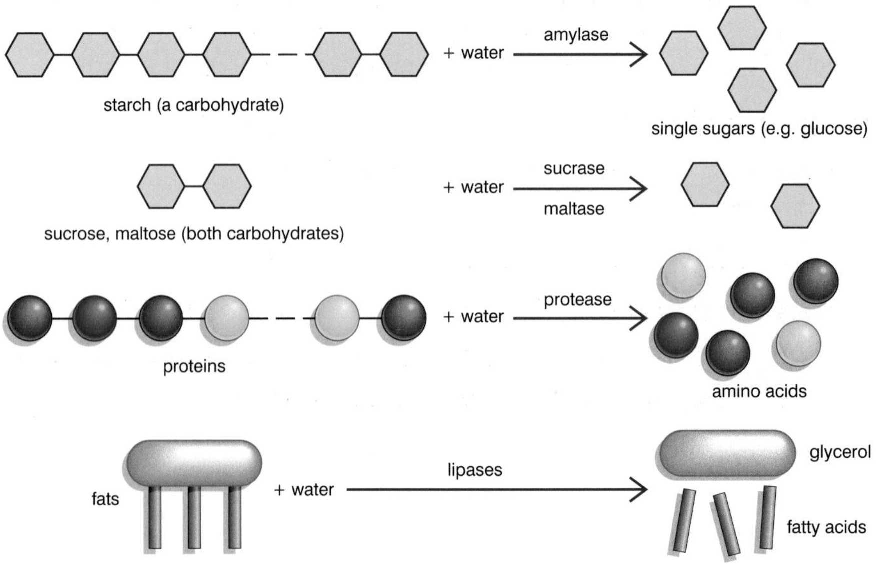
+ water
amylase
starch (a carbohydrate)
single sugars (e.g. glucose)
+ water
sucrase
maltase
sucrose, maltose (both carbohydrates)
+ water
protease
proteins
amino acids
fats
+ water
lipases
glycerol
fatty acids

Breaking down food for absorption by chemical digestion. The names of specific enzymes are shown above and below the blue arrows.

⌷ Enzymes

- Chemical digestion happens because of chemicals called **enzymes**. Enzymes are a type of **catalyst** found in living things.

FEATURES OF ENZYMES

- Enzymes are **proteins**.

- Enzymes are produced by cells.

- Enzymes help chemical substances change into new substances.

- Enzymes are **specific**, which means that each enzyme only works on one substance.

- Enzymes work best at a particular temperature (around 35–40°C for digestive enzymes), called their **optimum temperature**. At temperatures that are too high the structure of an enzyme will be changed so that it will not work. This is irreversible and the enzyme is said to be **denatured**.

- Enzymes work best at a particular pH, called their **optimum pH**. Extremes of (very high or very low) pH can also denature enzymes.

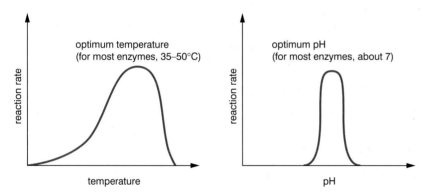

Enzymes work best at an optimum temperature and optimum pH.

GROUPS OF DIGESTIVE ENZYMES

- Every cell contains many enzymes, which control the many chemical reactions that happen inside it. **Digestive enzymes** are only one type of enzyme. They are produced in the cells lining parts of the digestive system and are **secreted** (produced) to mix with the food.

- There are different **groups** of digestive enzymes, such as:

  - **proteases** (produced by the stomach wall) – which break down proteins

  - **lipases** (produced by the pancreas) – which break down fats

  - **amylase** (produced by the salivary glands) – which breaks down starch

  - **maltase**, **sucrase** and **lactase** (produced by the small intestine) – which break down the sugars maltose, sucrose and lactose.

⌷ Substances that help digestion

■ **Hydrochloric acid** is secreted in the stomach. This is important to **kill bacteria** in food. Also, the enzymes in the stomach work best at a **low (acidic) pH.**

■ **Sodium hydrogencarbonate** is secreted from the pancreas to **neutralise the acid** leaving the stomach so that the enzymes in the small intestine can work.

■ **Bile** is produced in the liver, stored in the gall bladder, and passes along the bile duct into the duodenum. Bile **emulsifies fats**. It breaks down large fat droplets into smaller ones, which means that a larger surface area is exposed for chemical reactions to take place.

Bile lowers the surface tension of large droplets of fat so that they break up. This part of the digestive process is called emulsification.

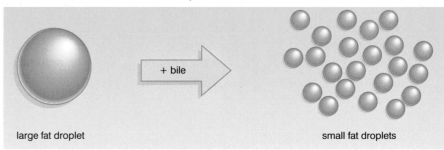

large fat droplet   + bile   small fat droplets

⌷ The digestive system

■ Eating food involves several different processes:

- **ingestion** – taking food into the body

- **digestion** of food into small molecules

- **absorption** of digested food into the blood

- **egestion** – removal of indigestible material (**faeces**) from the body.

■ Remember that egestion is not the same as **excretion**, which is the removal of waste substances that have been made in the body.

■ All these different processes take place in different parts of the **digestive system** (the **alimentary canal**).

■ Food moves along the digestive system because of the **contractions** of the muscles in the walls of the alimentary canal. This is called **peristalsis**.

Peristalsis moves food along the digestive system.

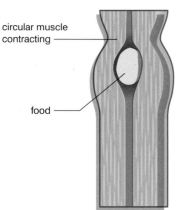

circular muscle contracting

food

movement of food

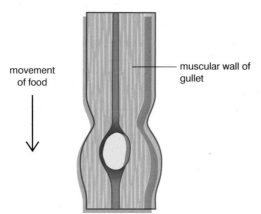

muscular wall of gullet

■ **Fibre** is an important part of the diet. Fibre is made up of the cell walls of plants. It adds bulk to food so that it can be easily moved along the digestive system by peristalsis. This is important in preventing constipation. Fibre is thought to help prevent bowel cancer.

Part of digestive system	What happens there
Mouth	Teeth and tongue break down food into smaller pieces Saliva from salivary glands moistens food so it is easily swallowed and contains amylase to begin breakdown of starch
Oesophagus or gullet	Each lump of swallowed food, called a **bolus**, is moved along by waves of muscle contraction called **peristalsis**
Stomach	Food enters through a ring of muscle known as a **sphincter** Acid and protease are secreted to start protein digestion Movements of the muscular wall churn up food into a liquid known as **chyme** (pronounced 'kime') The bulk of the food is stored while the partly digested food passes a little at a time through another sphincter into the duodenum
Gall bladder	Stores bile. The bile is passed along the bile duct into the duodenum
Pancreas	Secretes amylase, lipase and protease as well as sodium hydrogencarbonate into the duodenum
Small intestine (made up of duodenum and ileum)	Secretions from the gall bladder and pancreas as well as sucrase, maltase, lactase, protease and lipase from the wall of the duodenum complete digestion Digested food is absorbed into the blood through the villi
Large intestine or colon	Water is absorbed from the remaining material
Rectum	The remaining material (**faeces**), made up of indigestible food, dead cells from the lining of the alimentary canal and bacteria, is compacted and stored
Anus	Faeces is egested through a sphincter
Liver	Cells in the liver make bile. Amino acids not used for making proteins are converted into glycogen in the liver. Millions of red blood cells are broken down every day, and iron from their haemoglobin is stored in the liver. Vitamins A and D are stored in the liver. Poisonous compounds that are either produced by the body or enter the body are converted in to harmless substances. The liver removes excess glucose from the blood and stores it as glycogen.

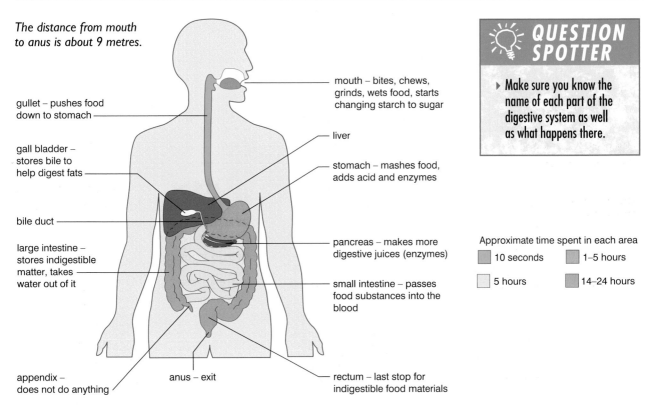

The distance from mouth to anus is about 9 metres.

gullet – pushes food down to stomach

gall bladder – stores bile to help digest fats

bile duct

large intestine – stores indigestible matter, takes water out of it

appendix – does not do anything

anus – exit

mouth – bites, chews, grinds, wets food, starts changing starch to sugar

liver

stomach – mashes food, adds acid and enzymes

pancreas – makes more digestive juices (enzymes)

small intestine – passes food substances into the blood

rectum – last stop for indigestible food materials

Approximate time spent in each area

	10 seconds		1–5 hours
	5 hours		14–24 hours

QUESTION SPOTTER

▶ Make sure you know the name of each part of the digestive system as well as what happens there.

⬚ Absorption of food

- After food has been digested it can enter the blood. This happens in the main part of the small intestine, known as the **ileum**. To help this process, the lining of the ileum is covered in millions of small finger-like projections called **villi** (singular: villus).

- The ileum is adapted for efficient absorption of food by having a large surface area. This is because:

  - it is **long** (6–7 metres in an adult)

  - the inside is covered with **villi**

  - the villi are covered in **microvilli**.

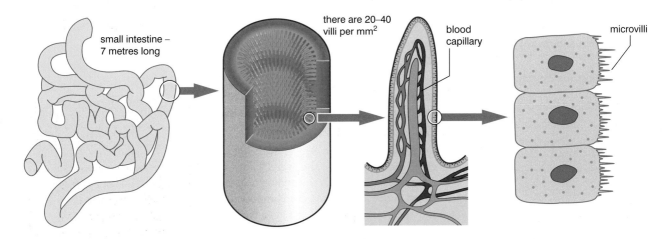

small intestine – 7 metres long

there are 20–40 villi per mm²

blood capillary

microvilli

The structure of the ileum.

- The villi themselves have features that also help absorption:

  - they have **thin, permeable walls**

  - they have a **good blood supply**, which maintains a concentration gradient that aids diffusion

  - they contain **lymph vessels** (lacteals), which absorb some of the fat.

- The lymph vessels eventually drain into the blood system.

? CHECK YOURSELF QUESTIONS

Q1 **a** Protein molecules are long chains of amino acid molecules joined together. Why do proteins need to be digested?
b What is the difference between physical and chemical digestion?
c What happens to enzymes at high temperatures?
d Why is the alkaline sodium hydrogencarbonate secreted into the duodenum?

Q2 **a** What is the difference between ingestion, egestion and excretion?
b Where do ingestion and egestion happen?

Q3 The villi are adapted to absorb food and the alveoli in the lungs are adapted to absorb oxygen. In what ways are they similar?

Answers are on page 138.

UNIT 3: ANIMAL ADAPTATIONS

 **Feeding adaptations**

❏ Digestive systems

- Organisms are **adapted** to their environment and this includes how they feed.

- Some animals produce enzymes that can break down **cellulose**. Mammals do not produce these enzymes. The digestive systems of mammals have adaptations for the diet they consume:

  - sheep and cows have a **rumen** between the oesophagus and stomach that contains bacteria to break down cellulose

  - rabbits have cellulose-digesting bacteria in a large **appendix**.

❏ Teeth

- **Herbivores** (plant-eating animals such as sheep) have teeth and jaws adapted for eating grass and vegetation:

  - **incisors** in the lower jaw are used to **cut** through grass by biting against a hard pad on the upper jaw

  - a flat surface on the **premolars** and **molars** is used to **grind** the vegetation

  - the jaws move from **side to side** to help grind the vegetation.

- **Carnivores** (meat-eating animals such as dogs) have teeth and jaws adapted for eating meat:

  - **carnassial** teeth and the premolars and molars are shaped to help **shear** meat and grind bones

  - sharp pointed incisors and canines are used to **tear** meat

  - the jaws have a strong **scissor** action because they only move up and down.

> ### ⚡ A* EXTRA
>
> ▸ The relationship between herbivores and cellulose-digesting bacteria is an example of mutualism. Herbivores benefit by obtaining sugar from the cellulose and the bacteria benefit by obtaining a supply of cellulose and nutrients.

> ### 💡 QUESTION SPOTTER
>
> ▸ You will **not** be expected to draw the structure of the tooth. You only need to be able describe how teeth are adapted to their function.

The skull and teeth of a human (left) and a dog (right).

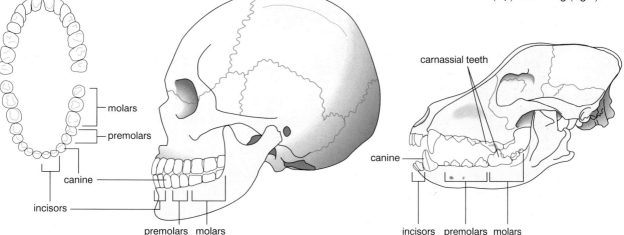

- Humans are **omnivores**: we have some teeth like herbivores and some teeth like carnivores. We can move our jaws from side to side and up and down. This means we can eat a wide range of foods.

⊡ Other adaptations

- Insects such as aphids, butterflies, houseflies and bees feed by sucking fluids into their mouths. They have a tube called a **proboscis** that can be uncurled and put into the food. The food is sucked up the proboscis. The proboscis of some insects, such as the mosquito, is sharp and needle-like so that it can penetrate skin.

- The diagram below shows some adaptations of an insect and its food source.

- The insect visits the flower to feed on nectar. However, the adaptations of the flower make sure that pollination takes place.

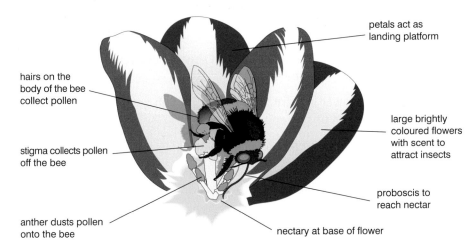

petals act as landing platform

hairs on the body of the bee collect pollen

large brightly coloured flowers with scent to attract insects

stigma collects pollen off the bee

proboscis to reach nectar

anther dusts pollen onto the bee

nectary at base of flower

QUESTION SPOTTER

▸ A question about malaria may also include ideas about how the mosquito is adapted to suck blood from an animal. Try to apply your knowledge of one topic to different situations.

- The larvae and adult stages of an insect are adapted to feed on different foods from different habitats. This means that each stage of the life cycle is not competing with the other for food.

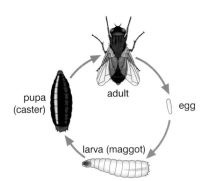

pupa (caster)

adult

egg

larva (maggot)

QUESTION SPOTTER

▸ In some countries maggots are added to wounds to help clean the area of infection. Questions often have a storyline that does not seem to link directly to what you have learnt. You will need to apply your knowledge to answer the question. In this case the question would be about the life cycle of an insect or adaptation of an animal for feeding.

- Some animals, such as mussels, are **filter feeders**. They feed on the **plankton** (microscopic organisms) that live in the water.

- Mussels feed by:

  • moving their cilia to draw water through their body

  • trapping plankton in their sieve-like gills

  • moving the trapped plankton to their mouth with more cilia.

water out

water in

shell

? CHECK YOURSELF QUESTIONS

Q1 Describe how insects, such as the housefly, are adapted for feeding.

Q2 Insects never seem to be short of a food supply. Explain why.

Q3 a Describe how the shape of the following teeth is adapted for its function:
  i incisors
  ii molars.
  b Suggest how the teeth of a lion would be different from the teeth of a sheep.

Answers are on page 139.

Adaptations for habitats

QUESTION SPOTTER

▸ You will be probably have to label a diagram of a joint. It is also important to be able to describe the function of the different parts.

⟲ Why do animals move?

■ Animals need to move to catch their food.

■ As well as having specialised adaptations for feeding, organisms also have special adaptations for moving through their surroundings. The movement of animals is called **locomotion**.

■ **Vertebrates** (animals with a backbone) have an **internal skeleton** made of **bone** that is used for:

- protection (of organs)

- support (to provide shape)

- movement (with the help of muscles).

■ Bones:

- are **rigid** – to provide support

- have **muscles** attached – to provide movement at the joints

- are **resistant** to compression, bending and stretching

- are **hardened** by deposits of calcium phosphate

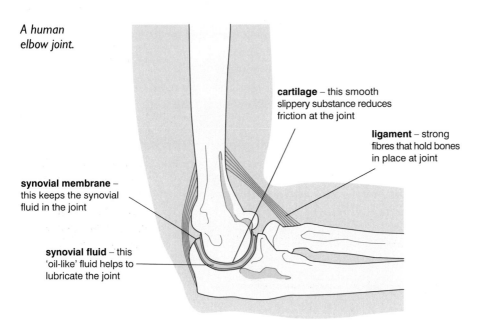

A human elbow joint.

cartilage – this smooth slippery substance reduces friction at the joint

ligament – strong fibres that hold bones in place at joint

synovial membrane – this keeps the synovial fluid in the joint

synovial fluid – this 'oil-like' fluid helps to lubricate the joint

- have **living** cells and proteins – to prevent them from being brittle.

■ **Muscle tissue** contains fibres that can contract when supplied with energy released from respiration. When muscular activity is increased more glucose and oxygen need to be supplied to the muscle tissue at a faster rate. Carbon dioxide and heat need to be removed from the muscle tissue.

■ **Tendons** attach muscles to bones.

EXCERCISING

■ Regular **exercise**:

- increases muscle strength

- keeps joints working smoothly

- keeps muscles toned

- improves circulation.

- Common **injuries** of a vertebrate body include:
  - dislocations – a bone is forced out of a joint
  - sprains – ligaments and tendons at a joint become torn.

HOW ARE BIRDS ADAPTED FOR FLIGHT?

- **Birds** are vertebrates that have special adaptations so that they can fly.

- Birds have **wings** that provide a large surface area.

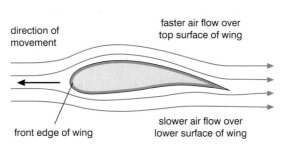

direction of movement

faster air flow over top surface of wing

front edge of wing

slower air flow over lower surface of wing

higher air pressure below pushes wing upwards

- As the wings push down the bird is lifted upwards and forwards.

- Birds have **feathers** that:
  - provide a large surface area
  - are light and strong
  - have interlocking barbs that ensure a smooth surface.

- During the upstroke of a wing, air is able to pass between the feathers because of the arrangement of the primary and secondary feathers.

- As well as wings and feathers, birds have other adaptations to help them fly:
  - **bones** with a **honey–comb structure** that are strong but light
  - a **streamlined** body shape that reduces air resistance
  - **strong breast muscles** to move the wings.

HOW ARE FISH ADAPTED FOR SWIMMING?

- **Fish** are vertebrates that have special adaptations so that they can **swim**.

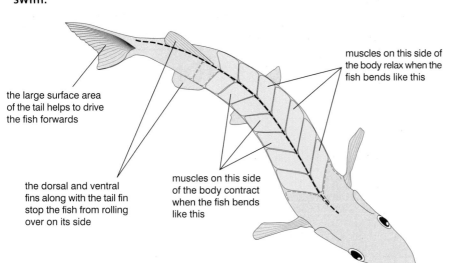

the large surface area of the tail helps to drive the fish forwards

the dorsal and ventral fins along with the tail fin stop the fish from rolling over on its side

muscles on this side of the body contract when the fish bends like this

muscles on this side of the body relax when the fish bends like this

Adaptations for the habitat

- **Fish** are so well adapted for living in water that they cannot live out of it.

- Fish obtain oxygen for respiration from water using **gills**.

  - Fish draw water through their mouths then pump the water over their gills so that **gaseous exchange** can take place.

  - Each gill is made up of many **filaments**, which contain tiny capillaries.

  - This number of filaments creates a **large surface area** for gas exchange.

  - As water flows over the gills oxygen **diffuses** from the water into the blood.

secondary lamellae

gill filaments

water flows out over the gills

water flows in through the mouth

gill rakers prevent damage to gills by filtering out tiny stones and other particles

gill bar supports the gills and carries blood vessels

gill arch

blood flow

blood and water flow in opposite directions

water flow

- **Amphibians** are also restricted to their habitat because of their method of gaseous exchange.

- Amphibians such as frogs have **thin** skin, which is always kept **wet**. This allows oxygen to **diffuse** through the skin.

- When active frogs need to draw air into their lungs, they move their throat up and down.

- **Reptiles** are more adapted for living on dry land. They have a scaly, waterproof skin and breathe with lungs.

CHECK YOURSELF QUESTIONS

Q1 a Which parts of a joint:
  i   attach muscles to bones
  ii act as shock-absorbers on the end of bones
  iii hold bones together at the joint?
b Explain why there are more joints in the fingers than the arm.

Q2 Explain how the curved shape of birds wings helps them to fly.

Q3 Explain how the following adaptations help fish to swim:
  a a zig-zag arrangement of muscles
  b large surface area of tail fins
  c streamlined body shape.

Answers are on page 139.

UNIT 4: HUMAN BODY MAINTENANCE

The nervous system

What does the nervous system do?

- The **nervous system** collects information about changes inside and outside the body, decides how the body should respond and controls that response.

Receptors

- Information is collected by **receptor cells**, which are usually grouped together in **sense organs**, also known as **receptors**.

- Each type of receptor is sensitive to a different kind of change or **stimulus**.

Sense organ	Sense	Stimulus
Skin	Touch	Pressure, pain, hot/cold temperatures
Tongue	Taste	Chemicals in food and drink
Nose	Smell	Chemicals in the air
Eyes	Sight	Light
Ears	Hearing	Sound
	Balance	Movement/position of head

The eye

- The **eye** is the receptor that detects light.

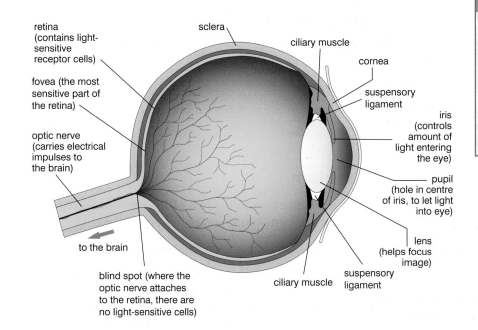

retina (contains light-sensitive receptor cells)

fovea (the most sensitive part of the retina)

optic nerve (carries electrical impulses to the brain)

to the brain

blind spot (where the optic nerve attaches to the retina, there are no light-sensitive cells)

sclera

ciliary muscle

cornea

suspensory ligament

iris (controls amount of light entering the eye)

pupil (hole in centre of iris, to let light into eye)

lens (helps focus image)

ciliary muscle

suspensory ligament

⚡ A* EXTRA

- There are two types of receptor cells in the retina, **rod cells** and **cone cells**. Rod cells work in dim light but cannot detect colour. Cone cells work in bright light and detect colour and fine details.

Part of eye	Job
Ciliary muscles	Contract or relax to alter the shape of the lens
Cornea	Transparent cover that does most of the bending of light
Iris	Alters the size of the pupil to control the amount of light entering the eye
Lens	Changes shape to focus light on to the retina
Humour	Clear jelly that fills the inside of the eye and provides shape for the eye ball
Retina	Contains light-sensitive receptor cells, which change the light image into electrical impulses There are two types of receptor cells: **rods**, which are sensitive in dim light but can only sense 'black and white', and **cones**, which are sensitive in bright light and can detect colour
Sclera	Protective, tough outer layer
Suspensory ligaments	Hold the lens in position
Optic nerve	Carries electrical impulses to the brain

■ The **iris** (ring-shaped, coloured part of the eye) controls the amount of light entering the eye by controlling the size of the hole in the centre, the **pupil**. The iris contains **circular** and **radial** muscles. In bright light the circular muscles contract and the radial muscles relax, making the pupil smaller. This reduces the amount of light entering the eye, as too much could do damage. The reverse happens in dim light, when the eye has to collect as much light as possible to see clearly.

■ The **lens** changes shape in order to focus light from objects that are distant or near.

The nervous system.

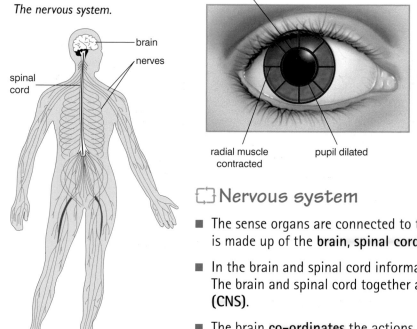

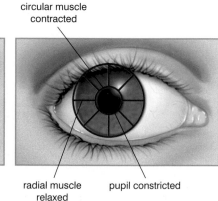

circular muscle relaxed

circular muscle contracted

radial muscle contracted   pupil dilated

radial muscle relaxed   pupil constricted

⬚ Nervous system

■ The sense organs are connected to the rest of the **nervous system**, which is made up of the **brain, spinal cord** and **peripheral nerves**.

■ In the brain and spinal cord information is processed and decisions made. The brain and spinal cord together are called the **central nervous system (CNS)**.

■ The brain **co-ordinates** the actions of the body.

■ Different areas of the brain are responsible for different actions.

Area of brain	Actions it controls
Cerebrum (cerebral hemispheres)	Thought, memory and intelligence Linking senses such as hearing and sight with muscles for voluntary response Feelings and emotions
Cerebellum	Sense of balance, muscular actions
Medulla	Reflex actions such as heart rate and breathing rate

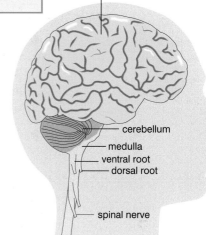

Areas of the brain.

- Signals are sent through the nervous system in the form of electro-chemical **impulses**.

TYPES OF NERVE CELLS (NEURONES)

- Nerve cells are called **neurones**.

- **Sensory neurones** carry signals to the CNS.

- **Motor neurones** carry signals from the CNS, controlling how the body responds.

- **Relay (intermediate or connecting) neurones** connect other neurones together.

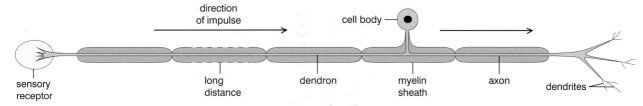

Sensory neurone

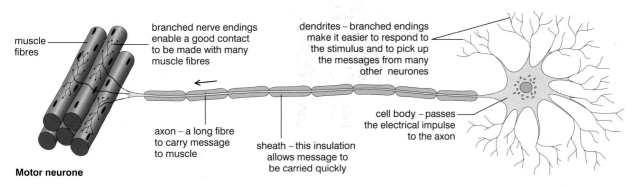

Motor neurone

Relay neurone

NEURONES ARE SPECIALISED

- Neurones can be **very long** to carry signals from one part of the body to another.

- Neurones have many branched nerve endings (**dendrites**) to collect and pass on signals.

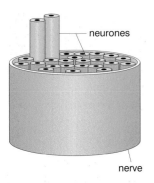

neurones

nerve

Cross-section of a nerve.

- Many neurones are wrapped in a layer of fat and protein, the **myelin sheath**, which insulates cells from each other and allows the impulses to travel faster.

- Neurones are usually grouped together in bundles called **nerves**.

- **Voluntary actions** always involve the brain. When you run to kick a ball your brain will involve the memory while forming a response.

⟲ Reflexes

- The different parts of the nervous system may all be involved when your body responds to a stimulus. The simplest type of response is a **reflex**. Reflexes are rapid, automatic responses that often act to protect you in some way, for example blinking if something gets in your eye or sneezing if you breathe in dust.

- The pathway that signals travel along during a reflex is called a **reflex arc**:

stimulus → receptor → sensory neurone → CNS → motor neurone → effector → response

- Simple reflexes are usually **spinal reflexes**, which means that the signals are processed by the spinal cord, not the brain. The spine sends a signal back to the **effector**. Effectors are the parts of the body that respond, either muscles or glands. Examples of spinal reflexes include standing on a pin or touching a hot object.

stand on pin → nerve endings → sensory neurone → spinal cord → motor neurone → leg muscles → leg moves

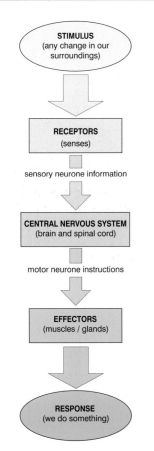

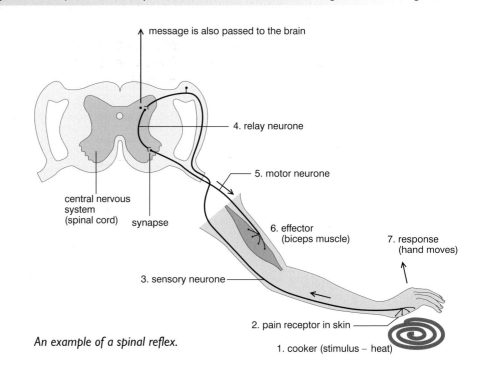

An example of a spinal reflex.

The flow of information from stimulus to response.

- When the spine sends a signal to an effector, other signals are sent on to the brain so that it is aware of what is happening.

- There are also reflexes in which the signals are sent straight to the brain. These are called **cranial reflexes**. Examples include blinking when dirt lands in your eye, or your pupil getting smaller as light is shone into your eye.

dorsal root → grey matter (centre of spinal cord) → ventral root

- There are also reflexes caused by a 'learnt' response. These are called **conditioned reflexes**. This type of reflex does not involve conscious thought. Examples include walking, food aversions and producing saliva at the smell of food.

smell of food → nose → sensory neurone → brain → motor neurone → salivary glands → mouth waters

Synapses

- Between nerve endings there are very small gaps called **synapses**, where signals are passed from one neurone to another even though the cells do not touch.

- Signals can be passed across synapses because, when an electrical impulse reaches a synapse, chemicals called **transmitter substances** are released from the membrane of one nerve ending and travel across to special **receptor sites** on the membrane of the next nerve ending, triggering off another nerve impulse.

- Sometimes signals are not passed across synapses, which stops us from responding to every single stimulus. (You can find out more about synapses in the section about drugs on pages 56–57.)

(You can find out more about synapses in the section about drugs on pages 56–57.)

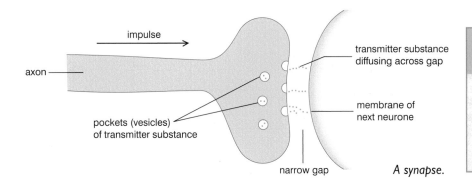

A synapse.

- impulse
- axon
- pockets (vesicles) of transmitter substance
- narrow gap
- transmitter substance diffusing across gap
- membrane of next neurone

CHECK YOURSELF QUESTIONS

Q1 a What is a stimulus?
b What is the difference between a receptor and an effector?

Q2 a What are the two parts of the CNS?
b What is the difference between a nerve and a neurone?
c Describe three features that neurones have in common with other cells.
d Describe two features that neurones have which make them different from other cells.

Q3 a Which of the following are reflexes?
A Coughing when something 'goes down the wrong way'.
B Changing channels on the TV when a programme ends.
C Jumping up when you sit on something sharp.
D Always buying the same brand of soft drink.
b Why are reflexes important?

Answers are on page 140.

The endocrine system

☐ Hormones

- **Hormones** are chemical messengers. They are made in **endocrine glands**.

- Endocrine glands do not have ducts (tubes) to carry away the hormones they make: the hormones are **secreted directly into the blood** to be carried around the body in the blood plasma. (There are other types of glands, called exocrine glands, such as salivary and sweat glands, that do have ducts.)

- Most hormones affect several parts of the body; others only affect one part of the body, called the **target organ**.

- The changes caused by hormones are usually slower and longer-lasting than the changes brought about by the nervous system.

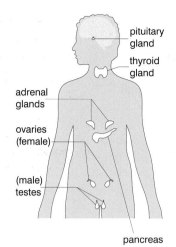

The endocrine system.

☐ What hormones are made where?

ADRENAL GLANDS

- The **adrenal glands** produce **adrenaline**. This is released in times of excitement, anger, fright or stress, and prepares the body for 'flight or fight': the crucial moments when an animal must instantly decide whether to attack or run for its life.

- The effects of adrenaline are:
  - increased heart rate
  - increased depth of breathing and breathing rate
  - increased sweating
  - hair standing on end (this makes a furry animal look larger but only gives humans goose bumps)
  - glucose released from liver and muscles
  - dilated pupils
  - paling of the skin as blood is redirected to muscles.

PITUITARY GLAND

- The **pituitary gland** produces **growth hormone, anti-diuretic hormone (ADH)** and some other hormones.

- Growth hormone helps mental and physical development in children.

- ADH instructs the kidney tubules to reabsorb more water into the blood.

- The pituitary gland also controls many other glands.

THYROID GLAND

- The **thyroid gland** produces **thyroxine**.

- Thyroxine helps mental and physical development in children.

⋆ IDEAS AND EVIDENCE

- You need to consider the power and limitations of science.

- The first 'test tube baby', Louise Brown, was born on 25 July 1978.

- The doctors had recorded the levels of pituitary hormones in the mother's blood. When these levels increased, they knew an egg was ripe and about to be released. The doctors removed the egg from the ovary before ovulation. The egg was fertilised outside the body and the embryo returned to the uterus.

▸ In exams you will not only be asked about the effects of hormones. You will also be asked to explain these effects.

▸ An example would be: 'Why is it important that adrenaline increases the heart rate in times of stress?'

PANCREAS

■ The **pancreas** secretes **insulin** and **glucagon**. (It also secretes digestive enzymes through the pancreatic duct into the duodenum.)

■ Insulin controls the **glucose level** in the blood. It is important that the blood glucose level remains as steady as possible. If it rises or falls too much you can become very ill.

■ After a meal, the level of glucose in the blood tends to rise. This causes the pancreas to release **insulin**, which travels in the blood to the liver. Here it causes any excess glucose to be converted to another carbohydrate, **glycogen**, which is insoluble and is stored in the liver.

■ Between meals, glucose in the blood is constantly being used up, so the level of glucose in the blood falls. When a low level of glucose is detected, the pancreas stops secreting insulin and secretes the hormone **glucagon** instead. Glucagon converts some of the stored glycogen back into glucose, which is released into the blood to raise the blood glucose level back to normal.

TESTES (MALES ONLY)

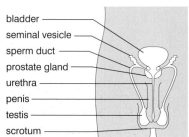

bladder
seminal vesicle
sperm duct
prostate gland
urethra
penis
testis
scrotum

The male reproductive system.

■ **Testosterone** (the male sex hormone) is secreted from the **testes**. Testosterone causes the following secondary sexual characteristics in boys:

- growth spurts
- hair growth on face and body
- penis, testes and scrotum growth and development
- sperm production
- voice breaking
- the body becoming broader and more muscular
- sexual 'drive' development.

OVARIES (FEMALES ONLY)

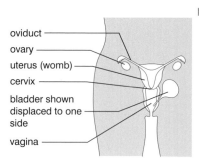

oviduct
ovary
uterus (womb)
cervix
bladder shown displaced to one side
vagina

The female reproductive system.

■ The **ovaries** produce the female sex hormones **progesterone** and **oestrogen**. These hormones cause the following secondary sexual characteristics in girls:

- growth spurts
- breast development
- vagina, oviducts and uterus development
- menstrual cycle (periods) starting
- hips widening
- pubic hair and under-arm hair growth
- sexual 'drive' development.

- These hormones also control the changes that occur during the **menstrual cycle**:
    - oestrogen encourages the repair of the uterus lining after bleeding
    - progesterone maintains the lining
    - oestrogen and progesterone control ovulation (egg release).

The menstrual cycle.

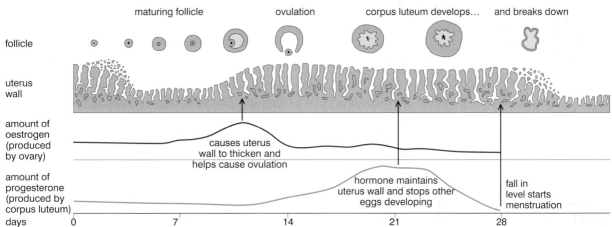

▢ The menstrual cycle

- About every 28 days an **egg** is released from one of a woman's ovaries. (The egg develops from one of the thousands of **follicles** present in the ovaries.) The egg travels down the **oviduct** to the **uterus** (womb) where, if it has been fertilised, it can implant and grow into a baby.

- To prepare for possible fertilisation, the lining of the uterus thickens. This is controlled by the release of progesterone from the **corpus luteum,** the remains of the follicle left behind in the ovary. If the egg has not been fertilised then the lining breaks down and is released (**menstruation**). Oestrogen from the ovary will encourage the uterus lining to grow again for the next egg released.

- If the egg has been fertilised, then progesterone continues to be released from the corpus luteum. This maintains the uterus lining during pregnancy and prevents further ovulation.

OTHER HORMONES INVOLVED IN THE MENSTRUAL CYCLE

- **Follicle-stimulating hormone (FSH)** is secreted by the pituitary gland. It causes eggs to mature in the ovaries and stimulates the ovaries to produce hormones such as oestrogen.

- **Luteinising hormone (LH)** is also secreted by the pituitary gland. It stimulates ovulation.

- The release of FSH and LH is partially controlled by oestrogen, which inhibits FSH production and stimulates LH production.

> ⚡ **A* EXTRA**
>
> ▸ The hormones involved in the menstrual cycle stimulate or inhibit each other's production. This is an example of a feedback system.

QUESTION SPOTTER

▸ An exam question may require you to write about information from several different units.
▸ An example would be: 'Describe the difference between nervous control and hormonal control of the body.'

⌑ Medical uses of hormones

■ Hormones can be used to treat various medical conditions, for example illnesses caused by a hormone not being made naturally in the correct quantities.

INSULIN

■ There are different types of diabetes, but in **diabetes mellitus** the body is unable to make enough of the hormone insulin. Insulin controls the level of glucose in the blood. Someone with diabetes cannot store excess glucose from their blood and it is excreted in urine instead. Other symptoms of diabetes include thirst, weakness, weight loss and coma.

■ Some people can control their diabetes with their diet and activity. For example, they make sure that they eat snacks between meals, eat some high-glucose food before any energetic activity and do not go for a long time without a meal.

■ Another way of controlling diabetes is to inject insulin before a meal.

■ Insulin can be extracted from animals' blood but is now produced by genetically engineered bacteria.

CONTROLLING FERTILITY

■ **Contraceptive pills** contain oestrogen and progesterone. They inhibit FSH production so that no eggs mature to be released.

■ FSH is used as a **fertility drug** to stimulate eggs to mature.

GROWTH HORMONE

■ **Injections of growth hormone** can be given during childhood to people that cannot produce enough of the hormone naturally, to ensure that growth is normal.

⌑ Illegal uses of hormones

■ Some drugs containing hormones, or chemicals similar to hormones, have been used by some athletes to **enhance their performance**. This is not only illegal but can have harmful side-effects.

CHECK YOURSELF QUESTIONS

Q1 Signals can be sent round the body by the nervous system and the endocrine system. How are the two systems different?

Q2 a How do the changes that adrenaline causes help the body react (fight or flight) to a stressful situation?
b After it has been released, adrenaline is broken down fairly quickly. Why is this important?

Q3 Explain the differences between glucose, glycogen and glucagon.

Answers are on page 141.

Homeostasis

☐ What is homeostasis?

■ For our cells to stay alive and work properly they need the conditions in and around them, such as the temperature and amount of water and other substances, to stay within acceptable limits. Keeping conditions within these limits is called **homeostasis**.

☐ Temperature control

■ The temperature inside your body is about 37°C, regardless of how hot or cold you may feel on the outside. This **core temperature** may naturally vary a little, but it never varies a lot unless you are ill.

■ Heat is constantly being **released** by respiration and other chemical reactions in the body, and is **transferred** to the surroundings outside the body. To maintain a constant body temperature these two processes have to balance. If the core temperature rises above or falls below 37°C various changes happen, mostly in the skin, to restore normal temperature.

Too cold?	Too hot?
Vasoconstriction: blood capillaries in the skin become narrower so they carry less blood close to the surface. Heat is kept inside the body.	**Vasodilation:** blood capillaries in the skin widen so they carry more blood close to the surface. Heat is transferred from the blood to the skin by **conduction**, then to the environment by **radiation**.
Sweating is reduced.	**Sweating:** sweat is released onto the skin surface and as it evaporates heat is taken away.
Hair erection: muscles make the hairs stand up, trapping a layer of air as insulation (air is a poor conductor of heat). This is more beneficial in animals but still occurs in humans (goose bumps).	Hairs lay flat so less air is trapped and more heat is transferred from the skin.
Shivering: the muscle action of shivering releases extra heat from the increased respiration.	No shivering.
A layer of fat under the skin acts as **insulation**.	

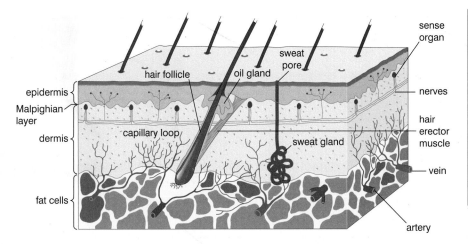

Section through skin.

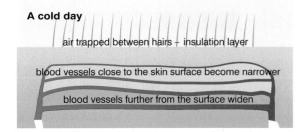

A cold day

air trapped between hairs – insulation layer

blood vessels close to the skin surface become narrower

blood vessels further from the surface widen

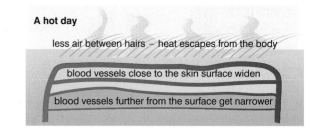

A hot day

less air between hairs – heat escapes from the body

blood vessels close to the skin surface widen

blood vessels further from the surface get narrower

- The core temperature is monitored by the **hypothalamus**: a part of the brain that monitors the temperature of the blood passing through it.

⌷ Water balance

- Our bodies are about two-thirds water. The average person loses and gains about 3 litres of water a day under normal conditions.

- Water is constantly being **lost** from the body:
  - in the air that is breathed out
  - in sweat (sweat also contains mineral salts)
  - in urine
  - in faeces.

- Water is **gained** by the body:
  - in food and drink
  - from respiration and other chemical reactions.

The water balance for an average person.

| minimum water gain | | minimum water loss |

food and drink: 1400ml

chemical reactions (e.g. respiration): 350ml

400ml from lungs when breathing out

500ml from skin as sweat

700ml from kidneys in urine

150ml from gut in faeces

total 1750ml
plus or minus optional extra of about 1000ml per day
Under dry and hot conditions, the body may
need up to an extra 9000ml per day

total 1750ml
plus or minus optional extra of about 1000ml per day

- **The water lost cannot be much more or less than the water gained.** This is why on a hot day, when you sweat more, you lose less water as urine, producing a smaller amount of more concentrated (darker) urine than normal.

THE ROLE OF THE KIDNEYS

- The amount of water that is lost from the body as urine is controlled by the **kidneys**.

- The kidneys regulate the amounts of water and salt in the body by controlling the amounts in the blood.

- The kidneys remove waste products such as **urea** from the blood. Urea is formed in the liver from the breakdown of excess amino acids in the body. This is an example of **excretion**.

- As blood flows around the body it passes through the kidneys, which remove urea and excess water and salt from the blood.

  1 Blood enters the kidneys through the **renal arteries**, which divide to form many **arterioles**. Each arteriole forms tiny capillaries that divide and coil, forming a **glomerulus**.

  2 A lot of the blood plasma (water, urea and other substances) is forced under great pressure into tiny tubules called **nephrons**. This process is called **ultrafiltration**.

  3 Further along each nephron, in the coiled tubules, useful contents (such as sugar) and some water are reabsorbed into the blood, leaving urea, excess water and excess salt. This mixture is called **urine**.

  4 More water is absorbed back into the blood, as needed by the body.

  5 The capillaries join up to form the **renal veins**, which carry blood away from the kidneys.

  6 The nephrons join up to form a **ureter**, which carries the urine to the bladder where it is stored until you go to the toilet, when the urine leaves the body through the **urethra**.

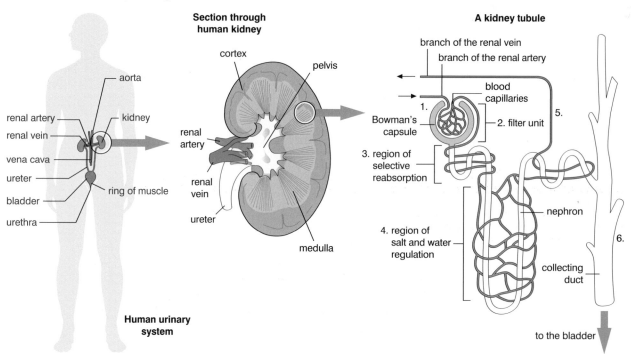

Human urinary system

Section through human kidney

A kidney tubule

■ You have two kidneys but you can survive with just one. However, if both kidneys become damaged through accident or disease a person may need to use a **dialysis machine**.

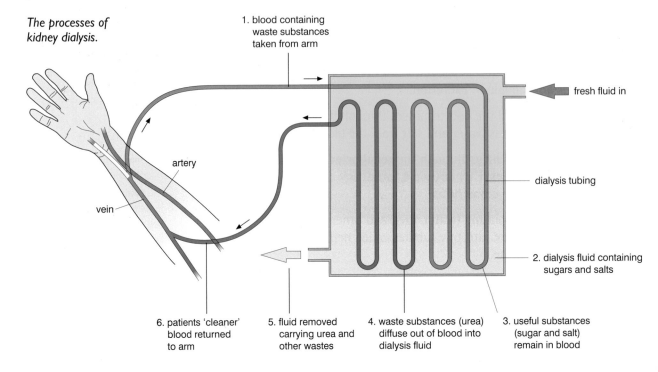

The processes of kidney dialysis.

1. blood containing waste substances taken from arm

fresh fluid in

artery

dialysis tubing

vein

2. dialysis fluid containing sugars and salts

6. patients 'cleaner' blood returned to arm

5. fluid removed carrying urea and other wastes

4. waste substances (urea) diffuse out of blood into dialysis fluid

3. useful substances (sugar and salt) remain in blood

MONITORING WATER BALANCE

■ Like temperature, the water content of the blood is monitored by the **hypothalamus**.

■ If there is too little water in the blood, for example if you have been sweating a lot, then the hypothalamus is stimulated and sends a hormone called **ADH** to the kidneys. This makes the kidneys reabsorb more water back into the blood so that less is lost in urine.

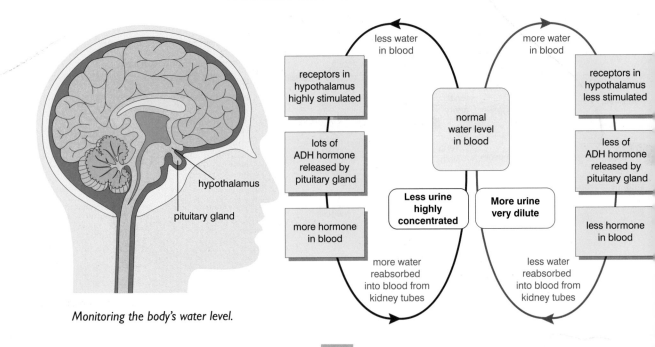

less water in blood

more water in blood

receptors in hypothalamus highly stimulated

normal water level in blood

receptors in hypothalamus less stimulated

lots of ADH hormone released by pituitary gland

less of ADH hormone released by pituitary gland

Less urine highly concentrated

More urine very dilute

more hormone in blood

less hormone in blood

hypothalamus

pituitary gland

more water reabsorbed into blood from kidney tubes

less water reabsorbed into blood from kidney tubes

Monitoring the body's water level.

- If your blood water level is high, for example if you have been drinking a lot, the hypothalamus is much less stimulated so less ADH is released, less water is reabsorbed and more water is lost in urine.

Negative feedback

- The body maintains a constant internal environment by monitoring changes and reacting to reduce these changes. This process is an example of **negative feedback**. It is 'negative' because the body attempts to **reduce** changes that occur.

EXAMPLES OF NEGATIVE FEEDBACK

- If the amount of glucose in your blood increases after a meal, insulin is released to reduce the amount.

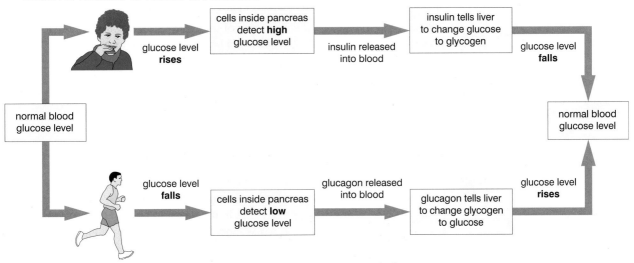

- If your body temperature falls, then shivering and vasoconstriction will increase the temperature.

- If you drink a lot, so increasing the amount of water in your blood, the kidneys will remove more water in urine.

- When you exercise you breathe faster. Muscles respiring more than usual use up oxygen faster. Faster breathing increases the amount of oxygen in the blood. These two processes counter each other, producing a steady oxygen level. The faster breathing also removes the extra carbon dioxide that has been produced by the muscles respiring. In fact, it is mainly the carbon dioxide levels in the blood that the brain monitors to control the rate of breathing. This is because high levels of carbon dioxide in the blood are toxic and have to be reduced quickly.

? CHECK YOURSELF QUESTIONS

Q1 A farmer who has been working outside all day on a hot summer's day produces a much smaller amount of urine than normal. It is also a much darker yellow colour than normal. Explain these points.

Q2 Why are the lungs organs of excretion?

Q3 How would the blood in the renal arteries be different from the blood in the renal veins?

Answers are on page 141.

Virus

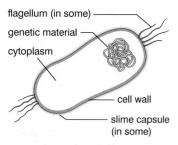

- protein coat
- genetic material

Bacterium

- flagellum (in some)
- genetic material
- cytoplasm
- cell wall
- slime capsule (in some)

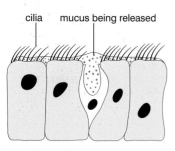

These two diagrams are not to scale. Viruses are much smaller than bacteria. They can only reproduce inside living cells.

cilia   mucus being released

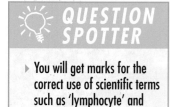

Cells lining the air passages in the trachea, bronchi and bronchioles.

QUESTION SPOTTER

▸ You will get marks for the correct use of scientific terms such as 'lymphocyte' and 'pathogen'.

Phagocytes

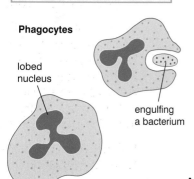

- lobed nucleus
- engulfing a bacterium

Preventing microbial attack

■ One main cause of disease is the presence of foreign organisms, **micro-organisms (microbes)**, in the body. Most of the time our bodies are able to prevent micro-organisms, such as **viruses** and **bacteria**, from getting in or from spreading. (You will find out more about disease and health in Unit 5.)

Barriers to infection	How they work
Skin	The skin provides a barrier that micro-organisms cannot penetrate unless through a wound or a natural opening
Mucus and ciliated cells in lining of respiratory tract	The lining of the nasal passages, and the trachea, bronchi and bronchioles in the lungs, is covered with a slimy mucus which traps air-borne micro-organisms as well as dirt. Tiny hair-like cilia use a waving motion to move the mucus upwards, where it usually ends up going down the oesophagus
Acid in the stomach	Stomach acid kills micro-organisms present in food and drink
Blood clots	At the site of a wound, blood platelets break open, triggering off a series of chemical changes that result in the formation of a network of threads at the wound. This network traps red blood cells, forming a clot. The clot prevents blood escaping and provides a barrier to infection

The immune system

■ If micro-organisms do manage to enter your body, they can cause harm by either damaging the cells around them or by releasing **toxins** (poisons) that make you ill. In each case they are attacked by **white blood cells**, which are part of our **immune system**.

■ There are two main types of white blood cells.

- **Phagocytes** that, because of their flexible shape, can engulf and then digest micro-organisms. This is called **phagocytosis**.

- **Lymphocytes** that produce chemicals called **antibodies**. Antibodies destroy micro-organisms in various ways, for example by killing the micro-organisms directly or by destroying the toxins they release (so acting as **antitoxins**).

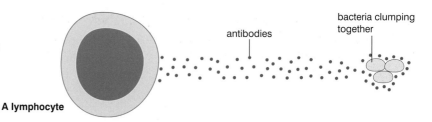

antibodies

bacteria clumping together

A lymphocyte

- Disease-causing micro-organisms (**pathogens**) have chemicals on their surface called **antigens**. If you catch a disease like chicken pox your lymphocytes make antibodies that stick to the antigens. The pathogens then stick together and they can be destroyed by the phagocytes.

antigens on surface of virus

antibodies released by white blood cell (lymphocyte)

antibodies attack the virus, sticking them together

⟳ Immunity

- Each pathogen has a unique antigen. So the lymphocytes produce different antibodies, specific to each pathogen. Once your body has made some antibodies, they remain in your blood for a long time. This means you are **immune** to the disease caused by that pathogen. This process of becoming immune is called **active immunity**.

- **Vaccination** causes active immunity. You are injected with a vaccine that contains a dead or mild form of a pathogen and your body makes the relevant antibody. Young children are given the MMR vaccine. This will help to protect them from measles, mumps and rubella.

- Sometimes you cannot wait for your body to make its own antibodies. If you cut yourself, you may be given an injection of ready-made tetanus antibodies. This is called **passive immunity**.

- The common cold virus can change the structure of its protein coat. You produce antibodies against the virus when you get a cold, but when the virus returns and has changed its protein coat the antibodies do not recognise the new structure. You are no longer immune and you get another cold.

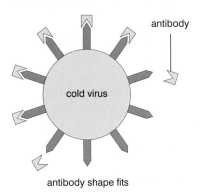
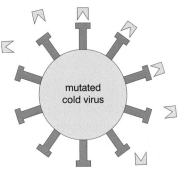

antibody

cold virus

antibody shape fits

mutated cold virus

antibody shape no longer fits

Viruses can change so that the antibody no longer affects it.

⟳ HIV

- **Human immunodeficiency virus (HIV)** attacks and invades white blood cells. The virus may then lay dormant for years. Gradually the virus instructs the white blood cells to make more copies of the virus. The number of white blood cells is reduced and the body is left open to attack from pathogens.

- As the number of white blood cells is reduced the immune system becomes less effective. **This leads to acquired immune deficiency syndrome (AIDS).** Patients with AIDS die from diseases their body is no longer able to prevent.

☆ IDEAS AND EVIDENCE

- Edward Jenner produced the world's first vaccine in 1796.

- Jenner noticed that milkmaids with cowpox never caught smallpox. He infected a young boy with cowpox. A few weeks later he injected the boy with smallpox. The boy did not develop smallpox.

- The result of Jenner's experimentation has saved many lives.

☆ IDEAS AND EVIDENCE

- Louis Pasteur developed a vaccine against rabies. He developed a weakened form of the virus.

- He injected the virus into an animal then took a tissue sample and allowed it to dry out. He then injected some of the tissue into another animal.

- He repeated this process until the virus was weak enough to inject into a person. After a few weeks he injected the person with a virulent form of the rabies virus, and the person did not develop rabies.

- Louis Pasteur's work has saved the lives of many people.

■ The spread of HIV can be prevented. People need to practise **safe sex** (by using a condom) and have fewer sexual partners. Drug users should not use dirty hypodermic needles.

Organ transplantation and blood groups

■ The reason that **organ transplants** are ideally donated by a close relative is that those organs are more likely to have **antigens** similar to the patient's own cells, and are therefore less likely to be attacked by the patient's immune system.

■ When a patient's immune system attacks a transplanted organ, the organ is said to be **rejected**. If the donor is not a close relative, **immunosuppressant drugs** have to be used to stop the transplant being rejected.

■ Before a transplant operation can take place a **donor** needs to be found. The donor needs to have the correct **tissue type**, one that matches the **recipient**. During the operation **blood transfusions** will be necessary, and the blood groups of the blood used must also match.

■ Everyone has a blood group that is either **A**, **B**, **AB** or **O**. The letters refer to the type of antigen on the surface of the red blood cells. People with blood group A have antigen A on their cell membranes and anti-B antibodies in their plasma.

Group	Antigen on cell membrane	Antibodies in plasma	Can donate blood to	Can receive blood from
A	A	Anti-B	A and AB	A and O
B	B	Anti-A	B and AB	B and O
AB	A and B	Neither	AB	All groups
O	Neither	Anti-A and anti-B	All groups	O

■ During a blood transfusion, if the two blood types do not match the donor's red cells are clumped in the patient's blood, which can lead to serious harm.

CHECK YOURSELF QUESTIONS

Q1 Edward Jenner produced the world's first vaccine in 1796.
  a Describe the process Jenner followed to develop the vaccine against smallpox.
  b Explain how a vaccine protects against a specific disease.
  c If you cut yourself, you may be given an injection to protect you from tetanus. Explain why this is a form of passive immunity.

Q2 The first case of HIV was identified in California in 1981. Its cause was identified in 1983.

  a Explain the difference between HIV and AIDS.
  b Suggest **three** ways that the spread of HIV can be prevented.
  c Patients with AIDS often die from pneumonia. Explain why.

Q3 **a** Most of the time there are very few white blood cells compared with the red blood cells in the blood. Explain why.
  b When would you expect the numbers of white blood cells to increase?

Answers are on page 142.

UNIT 5: HEALTH AND DISEASE

Types of disease

Causes of disease

- **Non–infectious diseases** can be caused by **environmental factors** (for example smoking) or by **genes** (inherited) (for example cystic fibrosis).

- Microbes cause **infectious diseases**. Microbes that cause disease are called **pathogens**. Pathogens are parasitic and can be bacteria, fungi, viruses and protozoa.

QUESTION SPOTTER

▸ In answers to questions about disease, do not use the word 'germs'. Use the word pathogen.

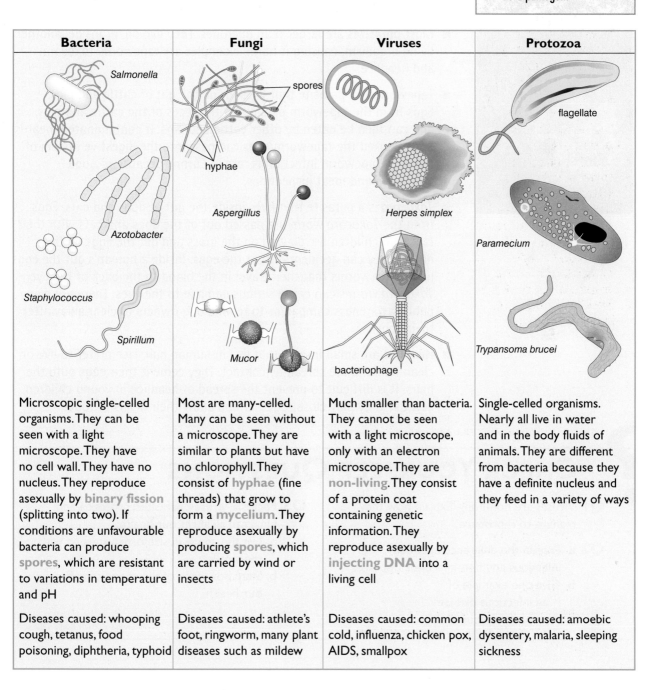

Bacteria	Fungi	Viruses	Protozoa
Salmonella	spores / hyphae / *Aspergillus* / *Mucor*	*Herpes simplex* / bacteriophage	flagellate / *Paramecium* / *Trypansoma brucei*
Azotobacter			
Staphylococcus			
Spirillum			
Microscopic single-celled organisms. They can be seen with a light microscope. They have no cell wall. They have no nucleus. They reproduce asexually by binary fission (splitting into two). If conditions are unfavourable bacteria can produce spores, which are resistant to variations in temperature and pH	Most are many-celled. Many can be seen without a microscope. They are similar to plants but have no chlorophyll. They consist of hyphae (fine threads) that grow to form a mycelium. They reproduce asexually by producing spores, which are carried by wind or insects	Much smaller than bacteria. They cannot be seen with a light microscope, only with an electron microscope. They are non-living. They consist of a protein coat containing genetic information. They reproduce asexually by injecting DNA into a living cell	Single-celled organisms. Nearly all live in water and in the body fluids of animals. They are different from bacteria because they have a definite nucleus and they feed in a variety of ways
Diseases caused: whooping cough, tetanus, food poisoning, diphtheria, typhoid	Diseases caused: athlete's foot, ringworm, many plant diseases such as mildew	Diseases caused: common cold, influenza, chicken pox, AIDS, smallpox	Diseases caused: amoebic dysentery, malaria, sleeping sickness

- Yeasts are an unusual family of fungi. Very few yeasts can form hyphae. The majority of yeasts:
  - are spherical cells
  - reproduce by budding
  - live where sugar is usually available (on the surface of fruits).

- The **protein coat** of a virus contains its genetic information (DNA or RNA). When a virus reaches a cell it injects its genetic material into the cell. The viral genetic material instructs the cell to stop its normal activity and make copies of the virus. When the cell has made hundreds of copies, it then dies. The new viruses are then released to go and invade other cells.

A virus attacking a cell.

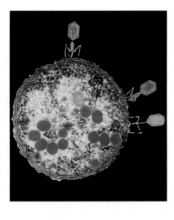

- Most **parasites** are larger than microbes. They live on, or inside, another living organism, causing it harm. Examples are tapeworm, toxocara and headlice.

- **Tapeworm** is a parasite that lives inside the gut of cattle and pigs. Eggs from the tapeworm pass out with faeces of the cattle and pigs, and can then be eaten by other cattle and pigs. If contaminated meat is undercooked the tapeworm eggs can pass into the digestive system of humans. Tapeworm infection is rare in Europe because of good sanitation and meat inspections.

- *Toxocara* is a parasite that lives inside the gut of dogs and cats. Eggs from the *Toxocara* worm are passed out of the dogs and cats with their faeces. If children are playing on the grass and get the eggs on their hands, they can accidentally eat the eggs. Inside a human's gut the eggs hatch into worms that then travel in the blood to the back of the eye. *Toxocara* worms can cause serious damage to the eyes. There are many public awareness campaigns to remind dog owners to clear away after their dogs.

- **Headlice** are small insects that live in human hair. Lice prefer to live on clean hair and are spread by contact. They cement their eggs onto the hairs. It is difficult to prevent the spread of headlice in young children, because they play close together and allow their heads to touch.

? CHECK YOURSELF QUESTIONS

Q1 Viruses are non-living. Explain how they manage to reproduce.

Q2 **a** Explain the difference between infectious and non-infectious diseases.
   b Give one example of:
   i   an infectious disease
   ii a non-infectious disease.

Q3 Bacteria, viruses, parasites, protozoa and fungi can all affect our health.
   a List the above in size order, starting with the smallest.
   b State how each of the above can affect our health.

Answers are on page 142.

The spread of disease

⊡ How do pathogens spread?

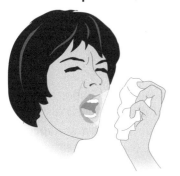

Sneezing and mouldy food can spread disease.

- Pathogens can be spread in various ways. A **vector** is something that can carry or spread disease.

Pathogen	How it is spread
Athlete's foot fungus	By contact
Hepatitis B virus	By body fluids
Influenza virus	In air
Salmonella bacterium	In food
Escherichia coli	In food
Cholera bacterium	In water
Dysentery bacterium	By animal vector (housefly)
Malarial protozoan	By animal vector (*Anopheles mosquito*)

> ⚡ **A* EXTRA**
>
> ▸ There is a danger of eating irradiated food in which the micro-organisms have been killed but their toxins remain.

- Simple hygiene can help to prevent the spread of infectious diseases like athlete's foot and the common cold.

Mosquitoes spread the malaria pathogen (see page 52).

- Within a local area, diseases can spread rapidly from person to person. This is known as an **endemic** outbreak of disease. However, with the modern increase of global travel, it is now common for diseases to be widespread, beyond local areas. Such a widespread outbreak of disease is called **epidemic**. This is one of the reasons why vaccination for travellers is so important.

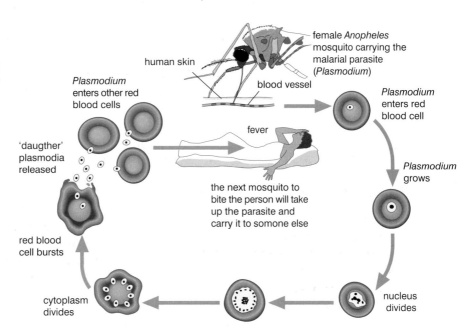

female *Anopheles* mosquito carrying the malarial parasite (*Plasmodium*)

human skin

blood vessel

Plasmodium enters other red blood cells

Plasmodium enters red blood cell

fever

Plasmodium grows

'daugther' plasmodia released

the next mosquito to bite the person will take up the parasite and carry it to somone else

red blood cell bursts

nucleus divides

cytoplasm divides

■ Animals spread many diseases. Bubonic plague, which was called the Black Death in the Middle Ages, is caused by bacteria carried from rats to humans by fleas.

■ The mosquito (*Anopheles*) carries the malarial parasite (*Plasmodium*) and feeds on human blood. When the mosquito feeds the malarial parasite enters the red blood cells and grows and multiplies. Its presence causes the human to suffer from malaria.

■ Water is used for drinking, washing and cooking. It is important to have clean water to reduce the spread of disease. In many underdeveloped countries disease is spread because **raw sewage** is released into rivers. In developed countries sewage is treated before it is released into rivers.

An example of a sewage treatment process.

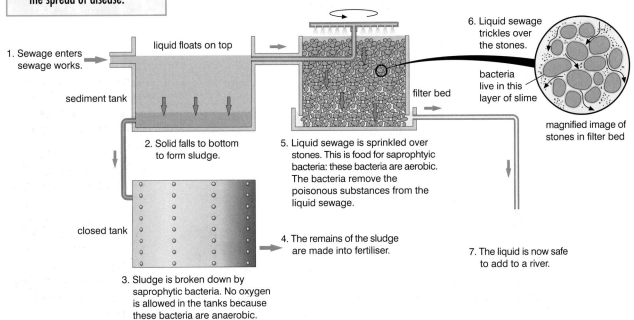

1. Sewage enters sewage works.

liquid floats on top

sediment tank

2. Solid falls to bottom to form sludge.

closed tank

3. Sludge is broken down by saprophytic bacteria. No oxygen is allowed in the tanks because these bacteria are anaerobic.

4. The remains of the sludge are made into fertiliser.

5. Liquid sewage is sprinkled over stones. This is food for saprophtyic bacteria: these bacteria are aerobic. The bacteria remove the poisonous substances from the liquid sewage.

filter bed

6. Liquid sewage trickles over the stones.

bacteria live in this layer of slime

magnified image of stones in filter bed

7. The liquid is now safe to add to a river.

? CHECK YOURSELF QUESTIONS

Q1 Public health awareness projects such as 'The Healthy Eating Campaign' are expensive to run. However, health authorities believe such projects may save money in the long term. Explain how public health awareness projects may save money for a health authority.

Q2 a What insect transmits malaria?
b Where inside the body does the malaria parasite reproduce?
c The malaria fever starts a few days after the insect bite. Explain why.

Answers are on page 143.

Stopping the spread of disease

⎕ Preventing food poisoning

- **Microbes** such as bacteria can reproduce very quickly. Some bacteria can reproduce every 20 minutes. For microbes to reproduce they need warmth, moisture and food.

- These conditions are found in kitchens. If a few bacteria get into food they can start to reproduce and after a few hours there can be thousands of bacteria.

- We need to **preserve** food to prevent food poisoning caused by microbes.

Method of preservation	Why it works
Freezing	Microbes stop reproducing at low temperatures
Pickling	Microbes cannot grow well in acid conditions, e.g. vinegar
Canning	Food is sterilised (microbes are killed) then sealed away from bacteria
Chemical preservation, including salt and sugar	Microbes cannot live in strong salt or sugar solutions. Water moves out of thebacteria into the surrounding solution by osmosis. The bacteria die due to lack of water. Chemical preservatives are also sprayed onto food to kill microbes
Drying	Microbes need water to live, so cannot live on dried food
Irradiation	Microbes can be killed without spoiling texture or taste of food

💡 QUESTION SPOTTER

▸ When answering questions about food preservation you need to say how the bacteria are affected by the process.

▸ When answering a question about salting or jam making, you need to explain how the high solute concentrations result in a decrease in water potential. It is not enough just to say 'by osmosis'.

✩ IDEAS AND EVIDENCE

▸ People are concerned about eating food that has been preserved by irradiation or with chemical preservatives.

▸ With irradiation, there is a slight risk that although the microbes have been destroyed their toxins are still present in the food. With preservatives, chemicals may accumulate in our bodies, although it may still be better to use them than to risk suffering or even dying from many kinds of food poisoning.

▸ You need to understand these issues and form your own opinions about them so that you can explain them in the exam.

✩ IDEAS AND EVIDENCE

▸ Louise Pasteur is known as the founder of microbiology. For hundreds of years people had believed that food just went bad and that maggots and insects in the bad food were just generated. After the discovery of the microscope some micro-organisms could be seen, but people still did not believe that they caused bad food or illness. Pasteur carried out a series of experiments to prove that micro-organisms were present in the air. He was able to demonstrate that it was the micro-organisms that caused food to go bad and illness.

▸ Pasteur sterilised some meat soup by boiling it. If the meat soup remained clear it was safe to eat, if it went cloudy micro-organisms had caused it to go bad. Pasteur made some glass flasks with a long S-shaped neck. The water in the S-shaped neck prevented micro-organisms from entering the flask. Pasteur left some flasks open to the air, some with S-shaped necks and some sealed. Only the flasks left open to the air contained meat soup that had gone bad.

▸ Pasteur's ideas of destroying micro-organisms by boiling are still used today. Milk is heated to 72°C for 15 seconds to destroy most of the bacteria. This is called **pasteurisation** and the process is referred to as the **flash** process.

▸ **Sterilisation** is also carried out with the use of chemicals such as chlorine.

▸ **Irradiation** is becoming a popular method of sterilising materials and food.

■ Intensive farming methods can result in the spread of *Salmonella* bacteria. Infected meat that is stored incorrectly, in warm conditions, means the *Salmonella* bacteria can reproduce quickly. The *Salmonella* produce toxins, which have harmful effects on the human digestive system and cause **food poisoning**. Good cooking techniques, correct food storage and improved animal rearing conditions help to reduce the number of food poisoning cases.

Bacteria on the surface of skin.

⟦⟧ Helping the immune system

■ Many microbes can live on the surface of our skin. Regular **washing** destroys some of these microbes.

■ If you cut your skin microbes can enter your body. **Antiseptics** are chemicals that are used to kill microbes. They prevent wounds becoming infected.

■ Antiseptics can be used on work surfaces, equipment and the skin. They cannot be taken internally.

✦ IDEAS AND EVIDENCE

▸ Joseph Lister discovered the use of antiseptics in the 1860s. At that time more than half of all patients died after an operation. Lister wanted to find a way to prevent his patients from dying after operations. He found that if a patient's wounds were sprayed with carbolic acid during the operation, they did not go septic.

▸ Lister's discovery greatly reduced the death rate in the late 1800s, and antiseptics are still very important in modern medicine.

■ **Antibiotics** are chemicals that can be taken internally, to destroy bacteria.

■ One of the most common groups of antibiotics is penicillins. These are produced by several genetic strains of the *Penicillium* fungus.

■ Penicillin works by infecting a wide range of bacteria and preventing them from developing a cell wall.

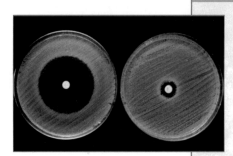

✦ IDEAS AND EVIDENCE

▸ Alexander Fleming discovered the first antibiotic in 1928. He left some bacterial cultures open to the air. Later he noticed some had been contaminated with a mould.

▸ He was about to throw them away, when he saw that where the mould was growing there was no longer any bacteria. The mould had killed the bacteria. This is shown in the picture.

▸ The work of Fleming and two other scientists, Florey and Chain, enabled the first antibiotic (penicillin) to be manufactured.

▸ The discovery and production of penicillin has saved millions of lives.

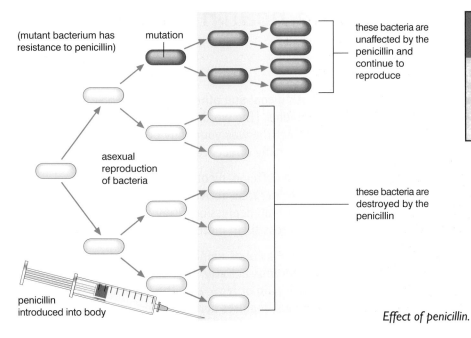

(mutant bacterium has resistance to penicillin)

mutation

these bacteria are unaffected by the penicillin and continue to reproduce

asexual reproduction of bacteria

these bacteria are destroyed by the penicillin

penicillin introduced into body

Effect of penicillin.

- Some bacteria are becoming **resistant** to a range of antibiotics. This means new antibodies constantly have to be developed.

- **Antibiotic resistance** can be reduced by using antibiotics less and making sure patients complete their course of antibiotics.

- You can take medicines to help relieve the symptoms of a disease. These medicines do not kill the pathogens.

A* EXTRA

▶ Phage technology is also used to overcome antibiotic resistance. Phages are a type of virus that attacks bacteria cells and destroys them. As bacteria change, so do the phages.

CHECK YOURSELF QUESTIONS

Q1 Jayne is suffering from a common cold.
  a Explain why antibiotics will not cure a common cold.
  b Explain how some bacteria have become resistant to a range of antibiotics.
  c Describe **two** methods used in an attempt to reduce antibiotic resistance.

Q2 Antiseptics and antibiotics are often used in hospitals.
  a Describe the difference between an antiseptic and an antibiotic.

  b Explain the difference between penicillin and *Penicillium*.
  c Which scientist's work enabled the first antibiotic to be manufactured?

Q3 Food can be preserved by several different methods.
  a Which methods of food preservation involve osmosis?
  b Explain how osmosis is used to preserve food.

Answers are on page 143.

Drugs and health

Drugs

■ The various drugs, both illegal and otherwise, that people may take can also cause health problems. (The word drug in this section means any substance that can affect the way your body behaves. The drug may have useful effects but many effects can also be dangerous.)

■ **Stimulants**, for example caffeine, nicotine, ecstasy, cocaine and amphetamines, **increase nervous activity**: they can speed up your heart rate, keep you awake or help depression. Stimulants affect synapses so that nervous impulses cross them more easily.

■ **Depressants**, for example alcohol, painkillers, tranquillisers, solvents and heroin, **decrease nervous activity**: they can slow your heart rate or reaction times, numb pain or relax you. They do this by restricting the movement of nervous impulses across synapses.

■ **Hallucinogens**, for example LSD and marijuana, affect the way you **perceive** things. People can see or hear things that are not there or see or hear things more vividly than normal.

■ Some drugs can be **addictive**: if you stop taking the drug you will show **withdrawal** symptoms, which may include cravings for the drug, nausea and sickness.

■ Another danger with drugs is that you build up a **tolerance** to them. This means that your body gets used to them and you have to take larger amounts of the drug for it to have the same effect.

TOBACCO

■ **Tobacco** contains **nicotine**, which is a stimulant that increases blood pressure and is addictive. Nicotine can also lead to the formation of blood clots, increasing the chances of heart disease.

■ Tobacco smoke contains **tar**, which irritates the lining of the air passages in the lungs, making them inflamed and causing **bronchitis**. Tar can cause the lining cells to multiply, leading to **lung cancer**. Tar also damages the cilia lining the air passages and causes **extra mucus** to be made, which trickles down into the lungs because the cilia can no longer remove it. Bacteria can breed in the mucus so you are more likely to get chest infections and smokers' cough as your body tries to get rid of the mucus. The **alveoli** are also damaged, so it is more difficult to absorb oxygen into the blood. This condition is known as **emphysema**.

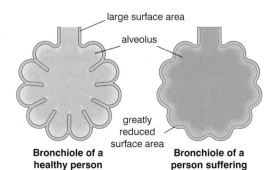

large surface area

alveolus

greatly reduced surface area

Bronchiole of a healthy person

Bronchiole of a person suffering from emphysema

- Tobacco smoke contains **carbon monoxide**, which stops red blood cells from carrying oxygen by combining irreversibly with the haemoglobin in the cells. This could also seriously affect the development of the fetus in the womb of a pregnant smoker, leading to a baby with low birth weight.

ALCOHOL

- **Alcohol** is a **depressant. Short-term effects** of alcohol include:

  - making you more relaxed, which is why in small amounts it can be pleasant

  - slower reactions and impaired judgement, which is why drinking and driving is so dangerous.

- Larger amounts of alcohol can cause lack of co-ordination, slurred speech, unconsciousness and even death.

- The **long-term use** of alcohol can cause **liver damage** (because the liver has the job of breaking down the alcohol), **heart disease**, **brain damage** and other dangerous conditions.

½ pint of beer (0.3 litre) 1 glass of wine 1 glass of sherry ½ pint of cider (0.3 litre) 1 single whisky

SOLVENTS

- **Solvents** are **depressants** and slow down brain activity.

- The effects last for a much shorter time than alcoholic effects but can include dizziness, loss of co-ordination and sometimes unconsciousness.

- Solvents can cause damage to the lungs, liver and brain.

- Death may also occur straight away because of heart failure.

THREAT OF HEART DISEASE

- Drinking large amounts of alcohol and smoking, combined with a diet containing a lot of fatty foods, little exercise and stress, can increase the chance of **coronary heart disease**. This is the main cause of death in Britain today. (You can read more about saturated fat and cholesterol in Unit 2.)

QUESTION SPOTTER

- Be aware of the dangers of passive smoking. Be ready for questions like: 'Should smoking be banned in public places?'
- Give reasons for your answer.

QUESTION SPOTTER

- If you are asked to describe the effects of alcohol, make sure you are clear whether the question is about its short-term or long-term effects.

CHECK YOURSELF QUESTIONS

Q1 Caffeine in coffee is a stimulant; alcohol is a depressant. Describe their effects on the human body.

Q2 Why is it dangerous to drive after drinking alcohol?

Answers are on page 144.

REVISION SESSION 1

▮ Photosynthesis ▮

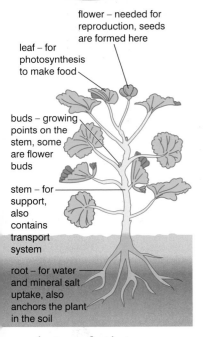

flower – needed for reproduction, seeds are formed here

leaf – for photosynthesis to make food

buds – growing points on the stem, some are flower buds

stem – for support, also contains transport system

root – for water and mineral salt uptake, also anchors the plant in the soil

Anatomy of a plant.

☐ What is photosynthesis?

■ Plants need and use the same types of foods (carbohydrates, proteins and fats) as animals but, while animals have to eat other things to get their food, plants **make it themselves**. The way they do this is called **photosynthesis**. The other ways that plants are different from animals, such as having leaves and roots, or being green, are all linked with photosynthesis.

■ In photosynthesis, plants take **carbon dioxide** from the air and **water** from the soil, and use the energy from **sunlight** to convert them into food. The first food they make is **glucose** but that can be changed later into other food types.

■ **Oxygen** is also produced in photosynthesis and, although some is used inside the plant for respiration (releasing energy from food), most is not needed and is given out as a **waste product** (although it is obviously vital for other living things).

■ The sunlight is absorbed by the green pigment **chlorophyll**.

■ The process of photosynthesis can be summarised in a **word equation**:

$$\text{carbon dioxide} + \text{water} \xrightarrow[\text{light}]{\text{chlorophyll}} \text{glucose} + \text{oxygen}$$

■ It can also be summarised as a balanced **symbol equation**:

$$6CO_2 + 6H_2O \xrightarrow[\text{light}]{\text{chlorophyll}} C_6H_{12}O_6 + 6O_2$$

QUESTION SPOTTER

▸ You will almost certainly be set at least one question on photosynthesis in your exams.
▸ The key to answering many questions about photosynthesis is remembering the equation.

■ Much of the glucose is converted into other substances, such as **starch**. Starch molecules are made of lots of glucose molecules joined together. Starch is insoluble and so can be stored in the leaf without affecting water movement into and out of cells by **osmosis**.

■ Some glucose is converted to **sucrose** (a type of sugar consisting of two glucose molecules joined together), which is still soluble, but not as reactive as glucose, so can easily be carried around the plant in solution.

■ The energy needed to build up sugars into larger molecules comes from respiration.

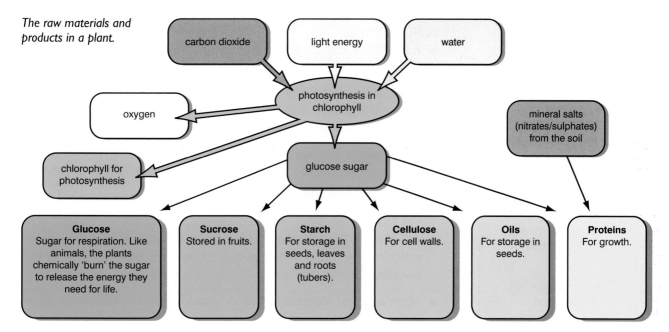

The raw materials and products in a plant.

Glucose Sugar for respiration. Like animals, the plants chemically 'burn' the sugar to release the energy they need for life.	**Sucrose** Stored in fruits.

(diagram contents:)

carbon dioxide — light energy — water → photosynthesis in chlorophyll → oxygen, chlorophyll for photosynthesis, glucose sugar

mineral salts (nitrates/sulphates) from the soil

Glucose Sugar for respiration. Like animals, the plants chemically 'burn' the sugar to release the energy they need for life.

Sucrose Stored in fruits.

Starch For storage in seeds, leaves and roots (tubers).

Cellulose For cell walls.

Oils For storage in seeds.

Proteins For growth.

⌖ Where does photosynthesis occur?

- Photosynthesis takes place mainly in the **leaves**, although it can occur in any cells that contain green chlorophyll. Leaves are adapted to make them very efficient as sites for photosynthesis.

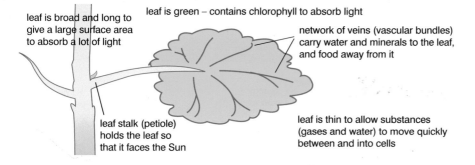

leaf is broad and long to give a large surface area to absorb a lot of light

leaf is green – contains chlorophyll to absorb light

network of veins (vascular bundles) carry water and minerals to the leaf, and food away from it

leaf stalk (petiole) holds the leaf so that it faces the Sun

leaf is thin to allow substances (gases and water) to move quickly between and into cells

- Leaves are **broad**, so as much light as possible can be absorbed.

- Each leaf is **thin**, so it is easy for carbon dioxide to diffuse in to reach the cells in the centre of the leaf.

- Leaves contain green **chlorophyll**, in the **chloroplasts**, which absorbs the light energy.

- Leaves have **veins** to bring up water from the roots and carry food to other parts of the plant.

- Each leaf has a stalk (**petiole**) that holds the leaf up at an angle so it can absorb as much light as possible.

INSIDE LEAVES

- A leaf has a transparent **epidermis** to allow light to travel to the cells within the leaf.

- There are many **palisade cells**, tightly packed together, in the uppermost half of the leaf, so that as many as possible can receive sunlight. Most photosynthesis takes place in these cells.

- **Chloroplasts** containing **chlorophyll** are concentrated in cells in the uppermost half of the leaf to absorb as much sunlight as possible.

- Air spaces in the **spongy mesophyll** layer allow the movement of gases (carbon dioxide and oxygen) through the leaf to and from cells.

- A leaf has a **large internal surface area to volume ratio** to allow the efficient absorption of carbon dioxide and removal of oxygen by the photosynthesising cells.

- Many pores or **stomata** (singular: stoma) allow the movement of gases into and out of the leaf.

- **Phloem** vessels transport products of photosynthesis away from the leaf

Cells of the leaf of a plant.

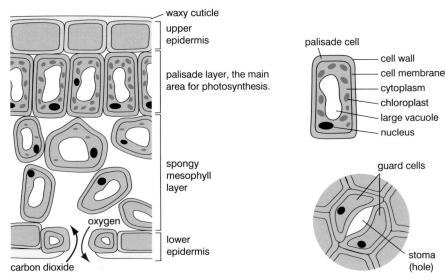

☐ Photosynthesis and respiration

- Once plants have made food using sunlight, they will at some stage need to release the energy in the food. They do this in the same way that humans and other animals do: **respiration**.

- During the day, plants respire 'slower' than they photosynthesise, so we only detect carbon dioxide entering and oxygen leaving the plant. During the night, photosynthesis stops and then we can detect oxygen entering and carbon dioxide leaving during respiration.

- At dawn and dusk, the rates of photosynthesis and respiration are the same and **no gases enter or leave** the plant because any oxygen produced by photosynthesis is immediately used up in respiration, and any carbon dioxide produced is used up in photosynthesis. These occasions are known as **compensation points**.

⊏⊐ Limiting factors

■ If a plant gets more light, carbon dioxide or water or a higher temperature, then it might be able to photosynthesise at a faster rate. However, the rate of photosynthesis will eventually reach a maximum because there is not enough of one of the factors needed: one of them becomes a **limiting factor**.

■ For example, if a farmer pumps extra **carbon dioxide** into a greenhouse the rate of photosynthesis might increase so the crop will grow faster. But if the **light** is not bright enough to allow the plants to use the carbon dioxide as quickly as it is supplied, the light intensity would be the limiting factor. The graphs show how the rate of photosynthesis is affected by limiting factors.

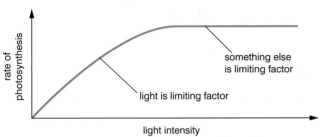
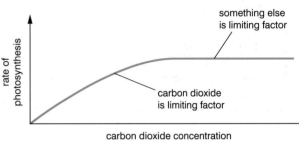

Increasing the levels of light and carbon dioxide, two of the factors necessary for photosynthesis, will increase the rate of photosynthesis until the rate is halted by some other limiting factor.

■ If the limiting factor in the first graph was the amount of carbon dioxide and the plants were then given more carbon dioxide, the graph would look like this:

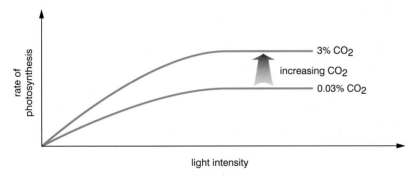

■ **Temperature** is also a limiting factor. Temperature affects the enzymes that control the rates of the chemical reactions of photosynthesis. Compare the shape of the graph on the right with the graph showing the effect of temperature on enzyme activity on page 21.

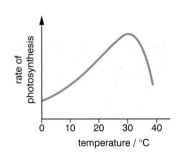

❓ CHECK YOURSELF QUESTIONS

Q1 How does a plant get the raw materials it needs for photosynthesis?

Q2 Why is the food that is made in the leaves not transported around the plant as starch?

Q3 Why are leaves usually broad and thin?

Answers are on page 144.

Transport in plants

☐ How are materials carried round a plant?

- In humans and many other animals, substances are transported around the body in the blood through blood vessels. In plants, water and dissolved substances are also transported through a series of tubes or vessels. There are two types of transport vessel in plants, called **xylem** and **phloem**.

- **Xylem vessels** are long tubes made of the hollow remains of dead cells. They carry **water** and **dissolved minerals** up from the roots, through the stem, to the leaves. They also give **support** to the plant.

- **Phloem vessels** are living cells. They carry **dissolved food materials**, mainly sucrose, from the leaves to other parts of the plant, for example growing roots or shoots or storage areas such as fruit. This movement of food materials is called **translocation**.

- In roots the xylem and phloem vessels are usually grouped together separately, but in the stem and leaves they are found together as **vascular bundles** or **veins**.

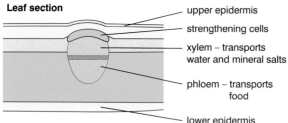

Leaf section
- upper epidermis
- strengthening cells
- xylem – transports water and mineral salts
- phloem – transports food
- lower epidermis

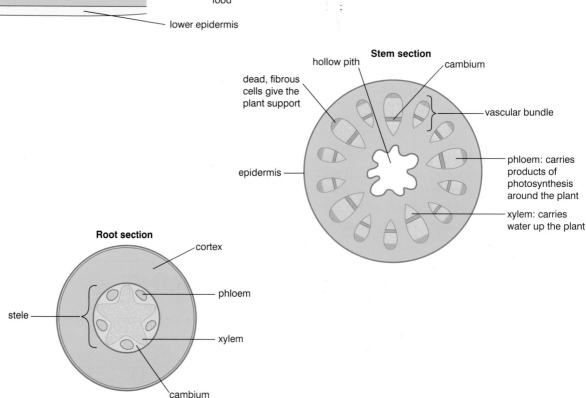

Stem section
- hollow pith
- dead, fibrous cells give the plant support
- cambium
- vascular bundle
- epidermis
- phloem: carries products of photosynthesis around the plant
- xylem: carries water up the plant

Root section
- cortex
- phloem
- stele
- xylem
- cambium

How do plants gain water?

- Roots are covered in tiny **root hair cells**, which increase the surface area for absorption. Water enters by **osmosis** because the solution inside the cells is more concentrated (has less water molecules) than the water in the soil.

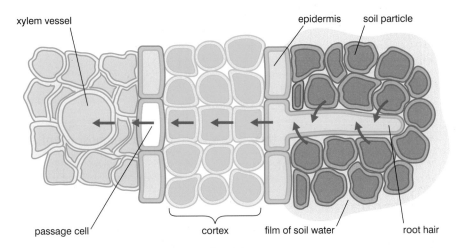

- Water continues to move between cells by osmosis until it reaches the **xylem** vessels, which carry it up to the leaves.

How do plants lose water?

- In the **leaves**, water moves out of the xylem and enters the leaf cells by osmosis (because the cells contain many dissolved substances). Water **evaporates** from the surface of the cells inside the leaf and then **diffuses** out through the open stomata. The evaporation of water causes more water to rise up the xylem from the roots rather like a drink flows up a straw when you suck at the top.

- Water loss from the leaves is known as **transpiration**. The flow of water through the plant from the roots to the leaves is known as the **transpiration stream**.

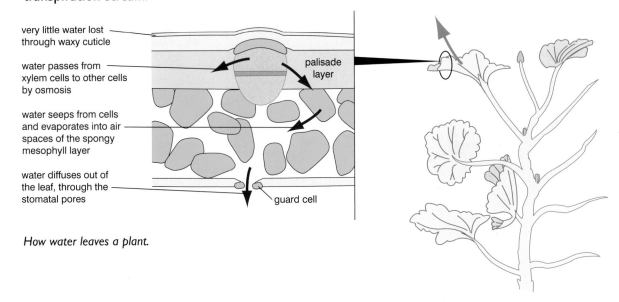

How water leaves a plant.

USES OF TRANSPIRATION TO PLANTS

- Transpiration brings up **water** and **minerals** from the soil.

- As the water evaporates it **cools** the plant.

SPEEDING UP TRANSPIRATION

- Transpiration happens faster in conditions that encourage evaporation, so when:
  - it is warm
  - it is windy
  - it is dry
  - it is very sunny (because this is when the stomata are most open)
  - the plant has a good water supply.

⟲ Water balance

- In a healthy plant the cells are full of water and the cytoplasm presses hard against the inelastic cell wall, making the cell **rigid**. The cell is **turgid** (has **turgor**).

- Most plants do not have woody tissue to hold them up: they are only upright because of cell turgor. If cells lose water, the vacuole shrinks and the cytoplasm stops pressing against the cell wall. The cell loses its rigidity and becomes **flaccid**. The plant will start to droop (**wilt**).

- If water loss continues the cytoplasm can shrink so much that it starts to come away from the cell wall. This process is known as **plasmolysis**.

- Plants need **open stomata** for the gases involved in photosynthesis to move in and out: but they also lose water through the stomata, and losing too much water can be a real problem. This is a particular problem for plants in places where water is not easily available, such as dry deserts or cold regions where the water in the soil is frozen.

Water plays a key role in the structure of plant cells.

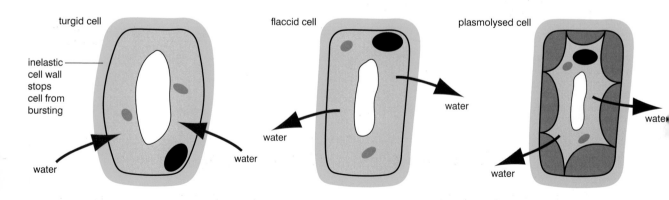

turgid cell

inelastic cell wall stops cell from bursting

water

water

flaccid cell

water

water

plasmolysed cell

water

water

water

WAYS OF REDUCING WATER LOSSES

- Leaves can have a **waxy cuticle**, so water cannot evaporate from the epidermal cells.

- Stomata are located mostly on the **underside** of leaves, where it is cooler and less evaporation occurs.

- Some stomata are **sunk** below the surface of the leaf so that they are less exposed.

- In some conditions a plant may **close** its stomata.

- Some plants have leaves covered with **hairs**. This creates a thin layer of still, moist air close to the leaf surface, which reduces evaporation.

- Plants in areas where water is very scarce reduce their leaves to **spines**, for example cactus spines and pine needles. This means the leaves have **less surface area**.

QUESTION SPOTTER

▸ Exams often have questions that ask you to describe several of the ways water loss from leaves can be reduced.

GUARD CELLS

- Around each stoma there are two **guard cells**.

- In **daylight** these cells absorb water by osmosis, which makes them **swell**. The inner walls are thickened and cannot stretch, unlike the outer ones. As the cells swell they curve and the gap between them opens up. During **darkness** the guard cells lose water by osmosis and the stomata **close**.

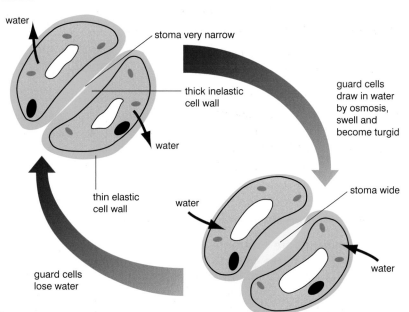

water

stoma very narrow

thick inelastic cell wall

water

guard cells draw in water by osmosis, swell and become turgid

thin elastic cell wall

water

stoma wide

water

guard cells lose water

The cell walls in guard cells are designed to open the stoma when the cell swells.

CHECK YOURSELF QUESTIONS

Q1 What are the differences between xylem and phloem?

Q2 Why does transpiration happen faster on a hot, sunny day than on a cool, dull day?

Q3 What is the difference between a turgid cell and a plasmolysed cell?

Answers are on page 145.

⚡ A* EXTRA

▸ Active transport allows roots to absorb minerals from very dilute solutions in the soil.

⊡ Why do plants need minerals?

■ To grow properly, plants need **minerals**, which they get from the **soil**. Some minerals are more important than others because they are needed in larger amounts.

Mineral	Element	Use in plant	Evidence of deficiency
Nitrates	Nitrogen	To make proteins, which are needed to make new cells	• Poor growth • Pale or yellow leaves
Phosphates	Phosphorus	Involved in respiration and photosynthesis	• Poor growth of roots and stems • Low fruit yield • Small purple leaves
Potassium salts	Potassium	To control salt balance in cells. Helps enzymes involved in respiration and photosynthesis	• Mottled leaves • Low fruit yield • Low disease resistance
Magnesium salts	Magnesium	The chlorophyll molecule contains magnesium	• Yellow patches between leaf veins

💡 QUESTION SPOTTER

▸ You may be asked about nitrates in questions about the nitrogen cycle, or in questions about fertilisers, as well as in questions about plants.
▸ Expect some of the questions in your exams to link together ideas from more than one part of your course.

⊡ Absorption

■ Minerals may only be present in the soil in low concentrations and so may need to be absorbed by the root hairs against a concentration gradient. This means that they have to be absorbed by **active transport** (see Unit 1).

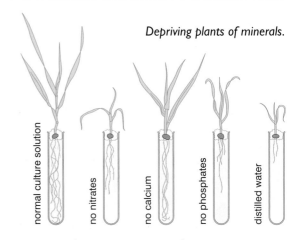

Depriving plants of minerals.

normal culture solution — no nitrates — no calcium — no phosphates — distilled water

■ Minerals enter the roots in solution (dissolved in water) and are carried in the **transpiration stream** through the xylem up the plant.

■ Although minerals are constantly being taken from the soil by plants, they are returned to the soil when animal and plant materials decay. (You will find out more about this and artificial fertilisers in Unit 7.)

? CHECK YOURSELF QUESTIONS

Q1 a Plants make proteins from carbohydrates. What extra elements do they need to do this?
b Why do plants need proteins?

Q2 Suggest why a lack of magnesium can cause yellow leaves.

Q3 Root hair cells contain many mitochondria. Suggest why.

Answers are on page 145.

Plant hormones

Controlling plant growth and development

- Many of the ways that plants grow and develop are controlled by chemicals called **plant hormones** or **plant growth regulators**.

- Like animal hormones, these chemicals are made in one part of the organism and travel to other parts where they have their effects. They are different from animal hormones in that they are not made in glands and obviously do not travel through blood: they **diffuse** through the plant.

What plant hormones control

- Plant hormones control:

  - the growth of roots, shoots and buds

  - flowering time

  - the formation and ripening of fruit

  - germination

  - leaf fall

  - healing of wounds.

Commercial uses of hormones

- **Selective weedkillers** kill the weeds in a lawn without harming the grass. The hormones make the weeds grow very quickly, which can then result in the death of the weed plants in a number of ways, for example the plant cannot support a quickly grown heavy structure, or the veins become too narrow for the effective movement of substances through the plant.

- **Rooting powder** is applied to plant cuttings to make them grow roots.

- Potatoes and cereals can be kept **dormant** so that they do not germinate during transport or storage.

- The **ripening** of soft fruit and vegetables can be **delayed** so that they are not damaged during transport.

- Crops such as apples can be made to **ripen at the same time** so that harvesting the crop is less time-consuming.

Tropisms

- **Tropisms** are **directional growth responses** to stimuli.

- Examples of tropisms include shoots growing towards the light and against the force of gravity, and roots growing downwards away from light but towards moisture and in the direction of the force of gravity. Growth in response to the direction of light is called **phototropism**. Growth in response to gravity is called **geotropism**.

QUESTION SPOTTER

▶ In your exam you may be asked to explain why particular plant hormones are put to particular uses by humans.

Commercial rooting powders can contain plant hormones.

A* EXTRA

▶ Auxin causes curvature in shoots by the elongation of existing cells, not by the production of more cells.

- Tropisms are controlled by a hormone called **auxin**. Auxin is made in the tips of shoots and roots and diffuses away from the tip before it affects growth. One effect of auxin is to inhibit the growth of side shoots. This is why a gardener who wants a plant to stop growing taller and become more bushy will take off the shoot tip, so removing a source of auxin.

- The growth of **shoots** towards light can be explained by the behaviour of auxin. Auxin moves **away** from the light side of a plant to the dark side. Its presence on the dark side encourages growth by increasing cell elongation. The dark side then grows more than the light side, and the shoot bends towards the light source.

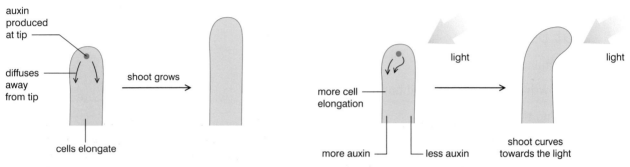

auxin produced at tip

diffuses away from tip

shoot grows

cells elongate

more cell elongation

light

light

more auxin — less auxin

shoot curves towards the light

IDEAS AND EVIDENCE

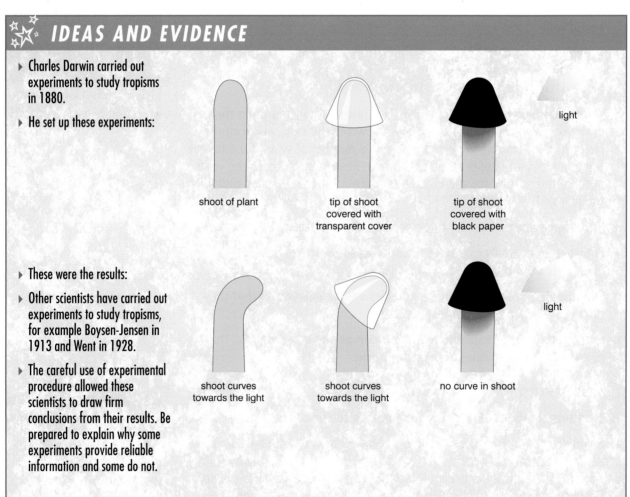

▸ Charles Darwin carried out experiments to study tropisms in 1880.

▸ He set up these experiments:

shoot of plant

tip of shoot covered with transparent cover

tip of shoot covered with black paper

light

▸ These were the results:

▸ Other scientists have carried out experiments to study tropisms, for example Boysen-Jensen in 1913 and Went in 1928.

▸ The careful use of experimental procedure allowed these scientists to draw firm conclusions from their results. Be prepared to explain why some experiments provide reliable information and some do not.

shoot curves towards the light

shoot curves towards the light

no curve in shoot

light

■ In roots, gravity causes the auxin to collect on the **lower** side. Here it **stops** the cells elongating, which causes the root to bend downwards.

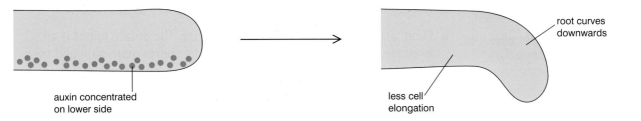

auxin concentrated
on lower side

root curves
downwards

less cell
elongation

CHECK YOURSELF QUESTIONS

Q1 How are plant hormones different from animal hormones?

Q2 Which of the following are controlled by plant hormones?
  a Flowering.
  b Pollination.
  c Germination.
  d Leaf fall.

Q3 Once a plant shoot has changed direction so it is growing towards the light, why does it stop bending?

Answers are on page 145.

REVISION SESSION 5

Plants and disease

QUESTION SPOTTER

▶ You will probably be asked to name an example of a plant disease.

Can plants get diseases?

■ Infectious diseases can also affect plants as well as animals.

■ Fungi are a frequent cause of plant disease. The potato blight is an example. Fungal spores land on leaves, the spores germinate and then enter the plant through the stomata. As the fungus spreads through the plant it destroys cells. Spores can lie dormant in the soil and attack the next crop.

■ Plant diseases can be controlled by various methods, for example:

 • **chemical treatment** – fungicides can be sprayed onto crops and soil

 • **selective breeding** – resistant varieties of plants can be developed

 • **crop rotation** – different crops can be sown in a field each year.

Potato blight is a fungus that causes plant disease.

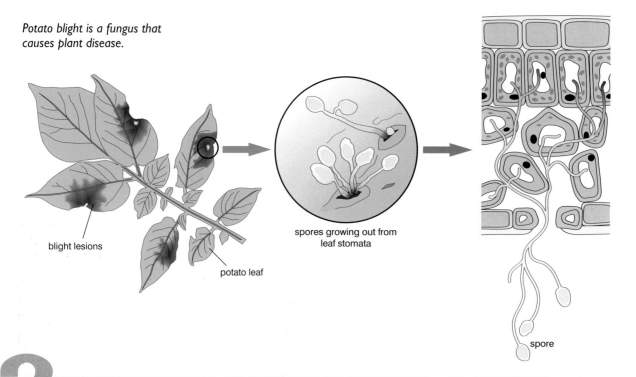

blight lesions

potato leaf

spores growing out from leaf stomata

spore

CHECK YOURSELF QUESTIONS

Q1 Potato blight is a fungus that attacks potatoes.
 a Describe the symptoms of this disesase.
 b List ways in which farmers try to control plant diseases.
 c Draw a sketch to show how the fungus infects the plant and causes the disease.

Q2 Describe how farmers could control plant diseases on their crops.

Q3 Genetic engineering has allowed scientists to develop varieties of plants that are resistant to most plant diseases. Give one reason for this genetic modification and one reason against.

Answers are on page 146.

UNIT 7: ECOLOGY
AND THE ENVIRONMENT

REVISION SESSION I  The study of ecology

Some ecological terms

- The environment is important to every living thing. We need to understand and protect our environment.

- **Ecology** is the study of how living things affect and are affected by other living things as well as by other factors in the environment.

- An **ecosystem** is a community of species and their environment, for example the species in a pond.

- **Environment** is a term often used to refer only to the physical features of an area, but can include living creatures.

- A **community** means the different species in an area, for example all the species in a grassland.

- A **species** is one type of organism, for example a robin. Only members of the same species can breed successfully with each other.

- **Population** means the numbers of a particular species in an area or ecosystem.

- An organism's **niche** is its way of life or part it plays in an ecosystem.

- An organism's **habitat** is where an organism lives.

Competition

- Animals and plants are always trying to survive and reproduce. However, there is always a 'struggle' for survival for various reasons.

- **Animals** struggle:
  - for food
  - for water
  - for space
  - for protection against the weather
  - against being eaten by predators
  - against disease
  - against accidents.

- Survival of **plants** is affected by:
  - the amount of water
  - the amount of light
  - space
  - the amount of minerals in the soil
  - weather
  - disease
  - being eaten.

Large numbers of birds have to compete with each other for resources such as space.

- In 1759 Thomas Austin introduced rabbits into Australia. He imported 24 rabbits from England and set them free on his property.

- The rabbits had very little competition and very few predators. The rabbit population increased quickly and they soon became a pest.

- In 1919 Dr Aragao realised the rabbits were a problem and suggested using the myxoma virus. The virus was tested for thirty years. In 1950 the virus was released into the environment in order to kill the rabbits.

- You need to be aware of the balance within an ecosystem. You should understand that when people change one organism in an ecosystem this has an effect on other organisms and conditions.

⚡ A* EXTRA

- Most predators and their prey don't show such a clear link as the lynx and hares because usually predators have more than one type of prey and animals are preyed on by more than one predator.
- This means that if the population size of one predator, or prey, falls then their place is taken by another.

- Animals and plants generally produce many young but their population sizes usually **do not vary significantly** from year to year because most of the young do not survive to adulthood. One reason for this is that the young have to **compete** for the resources they need, such as food, because there is not enough to go around.

- To help them survive, animals and plants are **adapted** to the environments in which they live. (You will find out more about adaptations in Unit 3.)

⬚ Predation

- One of the factors affecting a population of animals is the number of animals trying to eat them: their **predators**. The numbers of predators and **prey** are very closely connected.

- A famous example is that of snowshoe hares and lynxes in northern Canada. Both animals were hunted for their fur and a fur company kept records of how many animals were caught, so we can estimate the sizes of the populations for nearly a hundred years. This is such a vivid example because when a predator's prey reduces in number, the predator usually eats something else instead. In this case the lynx did not.

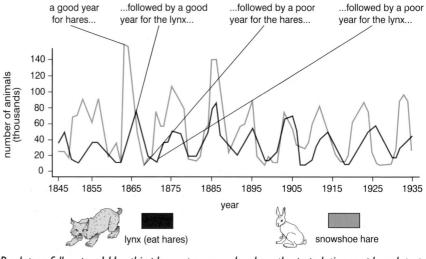

Predators follow prey! Use this phrase to remember how the population graphs relate to each other.

⬚ Human population

- The human population is increasing very rapidly in size. The graph of human population growth is getting steeper and steeper. Not only is the population increasing but the rate of increase is also increasing. This is an **exponential** increase.

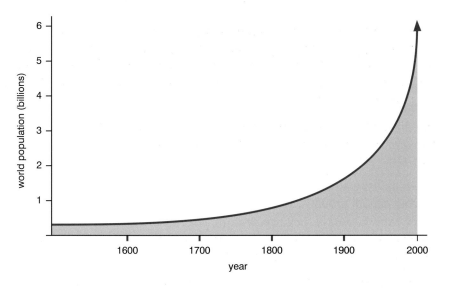

Reasons for the increase in the human population include:

- an increase in food production (through farming techniques)
- advances in medicine, such as immunisations and antibiotics
- improved living conditions.

Associations

- The way that organisms depend on each other for their life processes is known as **interdependence**.

- **Parasitism** is when **parasites** live in or on other organisms (their **hosts**) and harm the host. Examples are fleas living and feeding on a cat or tapeworms living inside a cow's gut.

- **Mutualism** is when organisms of different species live together very closely, to the benefit of both. For example, **nitrogen–fixing bacteria** live in the root nodules of **leguminous** plants such as peas, beans and clover. The bacteria get sugars, vitamins and a sheltered habitat from the plants. The bacteria provide the plants with the nitrogen-containing compounds they need to make proteins (see the nitrogen cycle on page 83).

QUESTION SPOTTER

- You will probably be asked to predict how the size of a population may change if other conditions change.
- Expect to be asked to explain your answer.

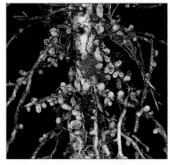

These root nodules contain nitrogen-fixing bacteria.

? CHECK YOURSELF QUESTIONS

Q1 In dense woodland there may not be many small plants growing on the ground. Suggest why.

Q2 In this country foxes eat rabbits. However, a graph of the two populations would not show the same oscillating (up and down) pattern as the lynxes and snowshoe hares on page 72. Why not?

Q3 In some summers there are much greater numbers of ladybirds than usual.
  a How would this affect the numbers of greenfly, their prey?
  b Why do the high numbers of ladybirds not persist year after year?

Answers are on page 146.

QUESTION SPOTTER

▶ These two pages describe some of the methods used for field studies. These methods will not be covered in your exam, but you will be taught them for practical work.

A simple pooter.

The standard way of using a pitfall trap.

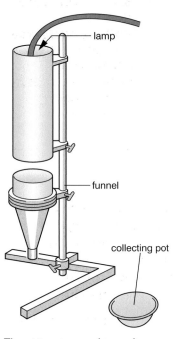

The apparatus used to make a Tullgren funnel.

Sampling

■ To monitor the population growth and distribution of small animals, sampling is carried out.

Ecological surveys

■ You may have already carried out an **ecological survey** of an area such as your school grounds. This involves finding out what organisms live there, where exactly they live, and the sizes of their populations. To do this you can use different equipment and techniques.

Pooter

■ The end of a short tube is placed over the animal and you suck on the other tube, collecting the animal in the bottle. The **pooter** is suitable for collecting small **insects** and **spiders**, which would otherwise quickly move away.

Pitfall trap

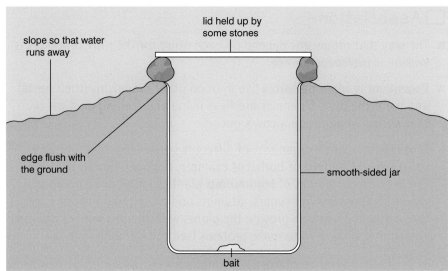

Pitfall trap (continued)

■ The **pitfall trap** can be left overnight. It is useful for collecting insects such as **beetles** that are more active at night and are too heavy to climb out. A lid or drainage holes will prevent water collecting and drowning the animals.

Tullgren funnel

■ Soil contains many **small animals**. One way of extracting these from the soil is to put a soil sample inside a **Tullgren funnel**. The heat from the lamp makes the animals go deeper until they fall through the sieve inside the funnel and can be collected in a pot underneath. The collecting pot can be filled with ethanol if you want to kill and preserve the animals.

Sweep net

- A **sweep net** can be waved or brushed over grass or other vegetation to collect any **insects** there.

Beating tray

- A **beating tray** is held under a bush or branch, which is then shaken to dislodge any **small animals** that can then be caught using a pooter.

Quadrat

- A **quadrat** can be used to carry out a **plant survey**, for example to compare the plant life in two different areas. The quadrat is placed randomly a number of times in each area, and each time the plants within the quadrat are identified and counted. You then combine the results to find the average for each area.

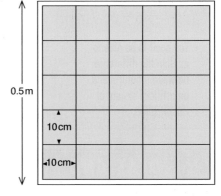

Sections of a quadrant.

- If the area of the region being investigated and the area of the quadrat are both known, then an estimate can be made of the total numbers of each species. Using quadrats in this way is called **sampling**, because only a sample of the region is inspected. Inspecting every plant throughout the region would take too long.

- Sampling using a quadrat works best if:

  - the quadrat is placed **randomly**

  - a **reasonable number of samples** is taken, so any 'odd' results will be 'evened out' when averages are taken

  - the plants being investigated are reasonably **evenly spread** throughout the area.

- You could also place a quadrat at regular intervals **along a line** to see how the habitat changes. This is called a **transect**.

- Quadrats can also be used to investigate the numbers or distribution of **animals** in a habitat, as long as they are evenly distributed and not constantly moving around the area.

⚡ A* EXTRA

- ▶ Using samples to work out the size of a population will only give you an estimate of the correct value.
- ▶ Your estimate will be more accurate the more samples you use.
- ▶ It will also be more accurate if the samples are taken randomly and are representative of the whole area being studied.

CHECK YOURSELF QUESTIONS

Q1 What equipment would be best to use to collect:
  a tiny insects living in the soil
  b large ground beetles that come out at night
  c small insects living in a thick bush
  d ants.

Q2 A pupil wanted to estimate the number of dandelions in a field. She used a large quadrat that was 1 m² and the field was 200 m². She used the quadrat 10 times and counted a total of 25 dandelions. Estimate the number of dandelions in the field.

Q3 Why is it important that quadrats are placed randomly?

Answers are on page 147.

Classification

⬚ Why *do we classify?*

- There are millions of different plants and animals in the world. To make **it easier to identify them**, we place them into different groups. The process of classifying organisms is called **taxonomy**.

- Classifying plants and animals helps us in our study of the environment.

CHANGING METHODS

- The first attempts to classify animals and plants were to group them according to a single characteristic, such as the ability to fly. So, for example, bats, wasps and owls could be placed in the same group. This method of classification is called an **artificial system**. An artificial system makes it difficult to make predictions about the grouped organisms.

- Modern scientists classify organisms using a **natural system**. This system groups organisms according to their most common characteristics, such as shape of leaf, number of legs, skin structure, skeleton.

☼ QUESTION SPOTTER

▸ You need to be able to explain the difference between a natural and an artificial system of classification.

✭ IDEAS AND EVIDENCE

▸ Our classification system has developed over many years.

▸ In Ancient Greece, Aristotle was the first person to carry out a scientific classification of living things. His system was used for nearly 2000 years.

▸ By the 17th century, people had began to travel great distances. More living organisms were being observed and recorded, and Aristotle's system was no longer adequate. John Ray (born in England in 1627) started to classify organisms according to many characteristics.

▸ Carl Linnaeus (born in Sweden two years after John Ray died) improved Ray's classification system. Linnaeus started the **binomial system** of classification.

- Living organisms are placed into groups, called **kingdoms**:

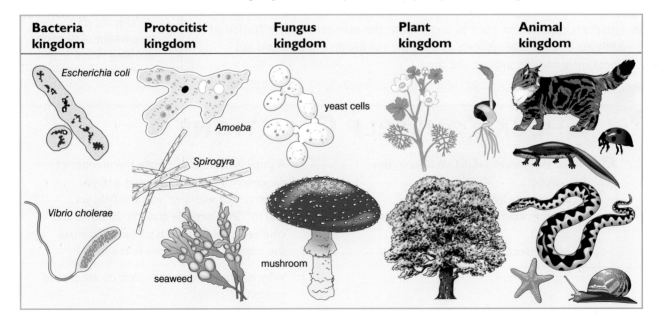

Bacteria kingdom	Protocitist kingdom	Fungus kingdom	Plant kingdom	Animal kingdom

Escherichia coli

Vibrio cholerae

Amoeba

Spirogyra

seaweed

yeast cells

mushroom

- The kingdoms are subdivided into **phyla**, then **classes**, **orders**, **families**, **genera** and **species**.

- The smallest classification group contains only one type of organism, called a **species**. The definition of a species is a group of similar organisms that can breed with each other to produce **fertile offspring**.

- Plants and animals are often called by different names in different parts of the country and world. The flower in the picture is often called a marsh marigold but can also be called a golden cup, king cup, may blob, horse blob and Mary bud. To avoid confusion, biologists use a standard international system for naming species. This is the **binomial system**.

- It is called the binomial system of naming because it gives every species a name made up of two parts. Each name is in Latin and is unique. The first part is the genus and the second part is the species. The correct biological name for the marsh marigold is *Caltha palustris*.

The marsh marigold, Latin name Caltha palustris.

☼ QUESTION SPOTTER

▸ You will **not** be expected to remember the Latin names of plants and animals.
▸ However, you will be expected to be able to explain that the first part is the genus name and the second part the species name.

CHECK YOURSELF QUESTIONS

Q1 It is not a good idea to classify plants only by the colour of their petals or the shape of their flowers. Explain why.

Q2 Describe the difference between an artificial system and a natural system of classification.

Q3 a What is meant by the scientific terms:
  i species
  ii binomial system.

b Explain why the binomial system is useful to biologists.

Answers are on page 147.

Relationships between organisms in an ecosystem

⟳ Food chains

- **Food chains** show how living things get their food. They also show how they get their **energy**. This is why they are sometimes written to include the Sun.

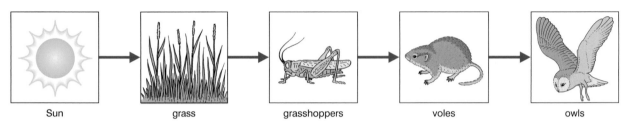

| Sun | grass | grasshoppers | voles | owls |

An example of a food chain.

QUESTION SPOTTER

▶ An example of a food chain may be given in the exam. Your answers should refer to the example provided.

- Food chains can be written going up or down a page, or even from right to left. All are correct, as long as the arrows always point **towards** the living thing that is taking in the food or energy.

STAGES OF A FOOD CHAIN

- **Producers** are the green plants (such as grass). They make their own food by **photosynthesis**, using energy from the Sun. (You can read more about photosynthesis in Unit 6.)

- **Primary consumers** (such as grasshoppers) are animals that **eat plants** or parts of plants, such as fruit. They are also called **herbivores**.

- **Secondary consumers** (such as voles) eat other animals. They may be called **carnivores** or **predators**.

- **Tertiary consumers** (such as owls) are animals that eat some secondary consumers. They are also called carnivores or predators.

- The animals hunted and eaten by predators are called **prey**.

- Animals that eat both plants and animals are called **omnivores**.

- The different stages of a food chain are sometimes called **trophic levels** (trophic means 'feeding'). So, for example, producers make up the first trophic level, primary consumers make up the second trophic level and so on.

- **Most food chains are not very long.** They usually end with a secondary consumer or a tertiary consumer, but occasionally there is an animal that feeds on tertiary consumers. This would be called a **quaternary consumer**.

Food webs

- **Food webs** are different food chains joined together. It would be unusual to find a food chain that was not part of a larger food web. Here is part of a food web:

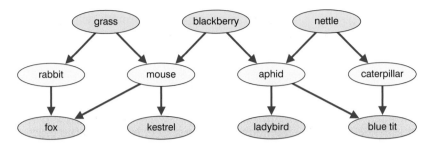

- Food webs show how different animals feed, and they can help us see what might happen if the food web is disturbed in some way.

- For example, what might happen if there was a disease that killed many, if not all, the rabbits? From the food web above you can see that the foxes would have to eat more of the other animals, which might result in fewer mice. On the other hand, there would be more grass for the mice to eat, which might mean there are more mice.

- It is not possible to say for sure what would happen to all the organisms in a food web because there are so many living things involved.

Pyramids

- **Pyramids of numbers** show the **relative numbers** of each type of living thing in a food chain or web by trophic level. Sometimes pyramids of numbers can be 'inverted', if, for example, the producers are much larger than the consumers.

A pyramid of numbers

An 'inverted' pyramid of numbers

- **Pyramids of biomass** show the mass of living material at each stage in a chain or web, so this shows what you would have if you could weigh all the producers together, then all the primary consumers and so on. Pyramids of biomass are almost **never** inverted.

- Pyramids either show organisms from a single food chain or the trophic levels in a food web. If they show trophic levels, **each stage of the pyramid may be labelled** as all the living things at a particular trophic level, for example producers or primary consumers.

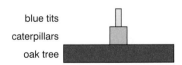

A pyramid of biomass

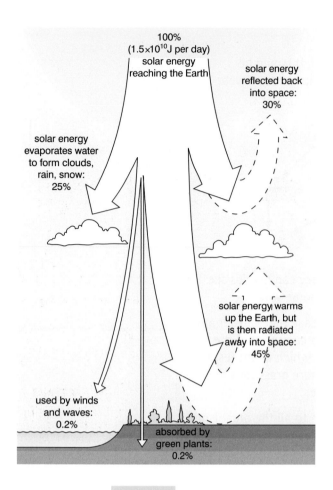

100%
(1.5×10^{10}J per day)
solar energy
reaching the Earth

solar energy
reflected back
into space:
30%

solar energy
evaporates water
to form clouds,
rain, snow:
25%

solar energy warms
up the Earth, but
is then radiated
away into space:
45%

used by winds
and waves:
0.2%

absorbed by
green plants:
0.2%

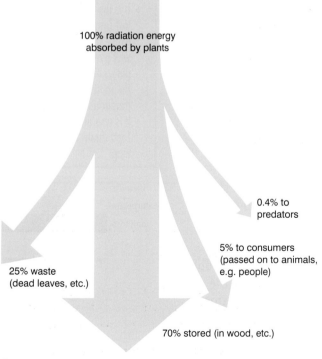

100% radiation energy
absorbed by plants

0.4% to
predators

5% to consumers
(passed on to animals,
e.g. people)

25% waste
(dead leaves, etc.)

70% stored (in wood, etc.)

Energy flow in plants.

⬡ Energy flow

- The arrows in food chains and webs show the **transfer of energy**. Not all the energy that enters an animal or plant is available to the next trophic level. Only energy that has resulted in an organism's growth will be available to the animal that eats it. For example, cows feeding on grass in a field:

Sun → grass → cow

- Only a **small proportion** of the energy in the sunlight falling on a field is used by the grass in photosynthesis:

  - some energy may be **absorbed** or **reflected** by **clouds** or **dust** in the air

  - some energy may be absorbed or reflected by **trees** or **other plants**

  - some energy may **miss** the grass and hit the ground

  - some energy will be reflected by the **grass** (grass reflects the green part of the spectrum and absorbs the red and blue wavelengths)

  - some energy will enter the grass but will **pass through** the leaves.

- The remainder of the energy can be used in photosynthesis.

- Some of the food made by the grass will be used for respiration to provide the grass's energy requirements, but some will be used for growth and will be available for a future consumer.

- However, not all the energy in the grass will be available to the cows:

  - some grass may be eaten by **other animals** (the farmer would probably call these pests)

  - the cows will **not eat all** of the grass (for example the roots will be left).

- Of the energy available in the grass the cows do **ingest** (eat), they will use only a small proportion for growth. The diagram on page 81 shows what happens to the energy in an animal's food.

HOW ANIMALS LOSE ENERGY

- Energy is lost from animals in two main ways:

1 **egestion** – the removal from the body, in faeces, of material that may contain energy but has not been digested

2 **respiration** – the release of energy from food is necessary for all the processes that go on in living things, most of this energy is eventually lost as heat. (You can read more about respiration in Unit 2.)

- Because **energy is lost** from the chain **at each stage** you almost always get a pyramid shape with a pyramid of biomass.

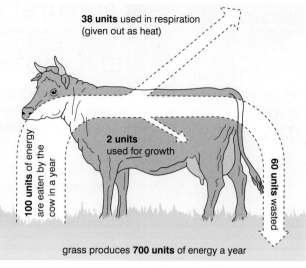

38 units used in respiration (given out as heat)

100 units of energy are eaten by the cow in a year

2 units used for growth

60 units wasted

grass produces **700 units** of energy a year

The energy flow in a young cow.

⬚ Farming

- A farmer raising plant or animal crops will get a **greater yield** by **reducing energy losses** from the food chain.

- For a **plant** crop this can be done by:
  - removing plants (weeds) that are **competing** with the crop for light
  - removing other organisms (pests) that might **damage** the plant crop
  - making sure the plants have enough **water** and **minerals** to be able to photosynthesise and grow efficiently.

- A farmer could keep **animals indoors** so they do not have to use as much energy to stay warm, or keep them **penned up** to reduce their movement and so lower the heat losses from respiration. This is sometimes known as **battery farming** and some people disagree with it on moral grounds.

- A lot of the crops we grow are for animal feed. Some people think that if we were all **vegetarians** we would need to grow fewer crops. There are fewer stages in the vegetarian food chain, so less energy is lost:

> Sun → plants → pigs → humans
>
> Sun → plants → humans

☼ QUESTION SPOTTER

▸ You will usually be required to describe some of the ways energy is lost from food chains.

? CHECK YOURSELF QUESTIONS

Q1 Look at this food chain in a garden:

rose bushes → aphids → ladybirds

   a Draw and label a pyramid of numbers for this food chain.
   b Draw and label a pyramid of biomass for this food chain.

Q2 Which is more energy efficient, to grow crops that we eat ourselves or to use those crops to feed animals which we then eat?

Q3 Why are there usually not more than five stages in a food chain?

Answers are on page 147.

Natural recycling

☐ Supplies are limited

- The **minerals** that plants need from the soil are mostly released from the **decayed remains** of animals and plants and their waste. This is one example of **natural recycling**. There is only a limited amount (on Earth) of the elements that living things need and use.

- Four of the most important elements in living things are **carbon (C)**, **hydrogen (H)**, **oxygen (O)** and **nitrogen (N)**. Important substances such as carbohydrates, fats and proteins are made up of carbon, hydrogen and oxygen. Proteins also contain nitrogen.

- The only way that animals and plants can continue to take in and use substances containing these elements is if the substances are constantly cycled around the ecosystem for reuse.

☐ The carbon cycle

QUESTION SPOTTER

▸ You may be given part of the carbon cycle or the nitrogen cycle and asked to add in the missing parts.

- Plants take in carbon dioxide because they need the **carbon** (and oxygen) to use in photosynthesis to make carbohydrates and then other substances such as protein.

- When animals eat plants they use some of the carbon-containing compounds to grow and some to release energy by respiration.

- As a waste product of respiration, animals breathe out carbon as carbon dioxide, which is then available for plants to use. (Don't forget that plants also respire producing carbon dioxide.)

- Carbon dioxide is also released when animal and plant remains decay (**decomposition**) and when wood, peat or fossil fuels are burnt (**combustion**).

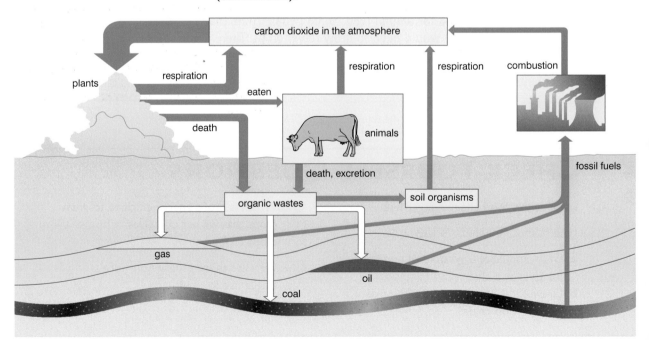

⌂ The nitrogen cycle

- Living things need **nitrogen** to make proteins, which are needed, for example, to make new cells for growth.

- Air is 79% nitrogen gas (N_2), but nitrogen gas is very unreactive and cannot be used by plants or animals. Instead plants use nitrogen in the form of **nitrates (NO_3^- ions)**.

- The process of getting nitrogen into this useful form is called **nitrogen fixation**.

- Decay bacter... down the rem... plants and animals, as well as animals' waste products, to produce ammonium compounds.
- Nitrifying bacteria convert the ammonium compounds to nitrates.

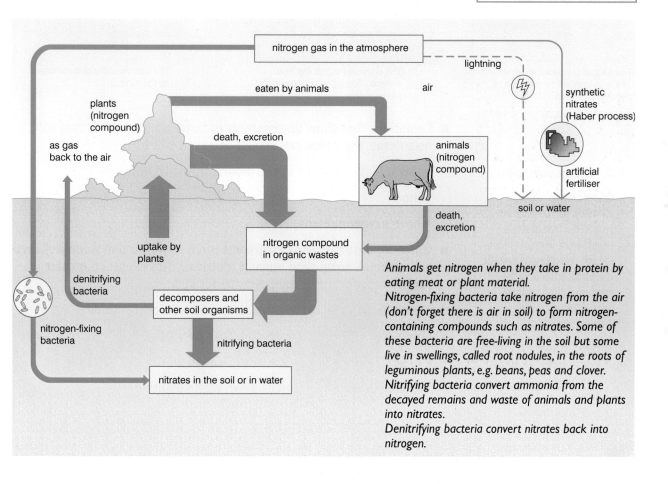

Animals get nitrogen when they take in protein by eating meat or plant material.

Nitrogen-fixing bacteria take nitrogen from the air (don't forget there is air in soil) to form nitrogen-containing compounds such as nitrates. Some of these bacteria are free-living in the soil but some live in swellings, called root nodules, in the roots of leguminous plants, e.g. beans, peas and clover.

Nitrifying bacteria convert ammonia from the decayed remains and waste of animals and plants into nitrates.

Denitrifying bacteria convert nitrates back into nitrogen.

⌂ Decomposers

- The **decay** of dead animal and plant remains is an important part of biological recycling. Decay happens because of the action of **bacteria** and **fungi**. These micro-organisms are known as **decomposers**.

- The bacteria and fungi use **enzymes** to digest the dead material.

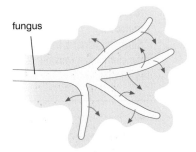

fungus

The fungus produces enzymes. The enzymes are released on to the dead material.

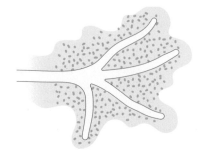

The enzymes digest the dead material. The digested material is now soluble.

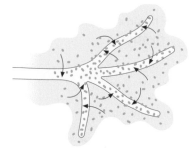

The soluble material is absorbed by the fungus.

- Conditions that allow decomposers to thrive are conditions that will help decay. The conditions should be:
  - **warm** (not so hot the micro-organisms would be killed)
  - **moist**
  - have **oxygen** present.

- Other organisms, called **detritivores** (such as worms and woodlice), feed on and break down dead remains (**detritus**). This exposes a greater surface area for the decomposers to act upon.

- Materials that can decompose are known as **biodegradable**.

- Micro-organisms that cause decay can be used by humans, for example:
  - in **sewage works**, to break down human waste
  - in **compost heaps**, to break down waste plant material.

❓ CHECK YOURSELF QUESTIONS

Q1 a In the carbon cycle, which main process removes carbon dioxide from the atmosphere?

b How is carbon dioxide put back into the atmosphere?

Q2 Nitrogen-fixing bacteria cannot live in waterlogged soil but denitrifying bacteria can, which is why this kind of soil is low in nitrates. Some plants that live in these conditions are carnivorous, trapping and digesting insects. Suggest why you often find carnivorous plants in bogs.

Q3 What is the difference between nitrifying and denitrifying bacteria?

Answers are on page 148.

REVISION SESSION 1 — Using microbes

How can we use microbes?

- As technology advances, we are turning our attention towards the effective use of different organisms.

- We use the properties and processes of some microbes to produce types of food and drink:

  - bread – a yeast (fungus) and sugar mixture are added to flour to form a dough, and the **respiration** of yeast produces carbon dioxide, which raises the dough

  - yoghurt – milk is pasteurised to kill any unwanted bacteria then the bacteria *Lactobacillus* is added and **cultured** at 46°C, releasing lactic acid into the milk to turn it into yoghurt

  - cheese – a culture of bacteria is added to warm milk and the resulting curds are separated from the liquid whey, more bacteria and special moulds are added to the curds and the cheese slowly ripens, with other moulds sometimes added to give flavour and colour to the cheese

  - vinegar – a mixture of beer, wine or ethanol and nutrients is poured over wood shavings coated with a bacteria *Acetobacter*, and this bacteria **converts** ethanol into vinegar

  - soy sauce – cooked soya beans and roasted wheat are fermented, filtered and then pasteurised

  - **single–cell protein (SCP)**, also called **mycoprotein** – this is a fungus that is mixed with carbohydrate and kept in warm conditions so that it grows rapidly, then the fungus is separated and dried

  - alcohol – the **anaerobic respiration** of yeast produces carbon dioxide and ethanol, which is the basis of alcohol, this process is called **fermentation**.

- In **wine–making**:

  - yeasts live on the surface of the fruit, usually grapes

  - natural sugars on the surface of the fruit is used by the yeast for respiration, resulting in fermentation.

- In **beer–brewing**:

  - inside germinating barley grains, the starch breaks down into a sugar solution, which is called the malting stage

  - the sugar solution is drained off and fermented with yeast

  - the flavour of the beer is developed by adding hops to the mixture.

QUESTION SPOTTER

▶ You will not be expected to remember the names of the different bacteria used for different food production processes, but you do need to be able to explain the different processes.

IDEAS AND EVIDENCE

▶ Where food or space is scarce, producing SCP can be a very good way of providing large quantities of a nutritious food source within a small area and using waste products from other processes. The process is expensive to set up, but once established can be cost-effective in the long term.

▶ You need to understand the value of SCP as a food source in terms of nutrition and economics.

- SCP is used as cattle food. It is also used to produce a **meat–free protein substitute** called Quorn. SCP has a very high level of protein and its production is very efficient. Compared with a conventional source of protein, for example from cattle, the production of SCP uses far less space and is much quicker. The process can be used to produce more protein from a food that is usually thrown away, for example SCP can be added to the whey in cheese production to produce a meat-free protein.

⌦ Using microbes on a large scale

- Microbes can be used to make chemicals on a **large scale**. For example, by growing the fungus *Penicillium* in a **fermenter** the antibiotic penicillin can be made in large quantities. The fungus is fed glucose and ammonia, and the fermenter is kept at 25°C.

- The **enzymes** that help to speed up the process of **fermentation** have an **optimum temperature**, for example of 25°C. Enzymes are sensitive to temperature. Most enzymes are denatured above 60°C and at low temperatures the enzymes work very slowly.

- A fermenter is filled with a small amount of the desired microbe, for example yeast, and a **growth medium**. The most suitable conditions for growth of the microbe are created.

Industrial fermenter

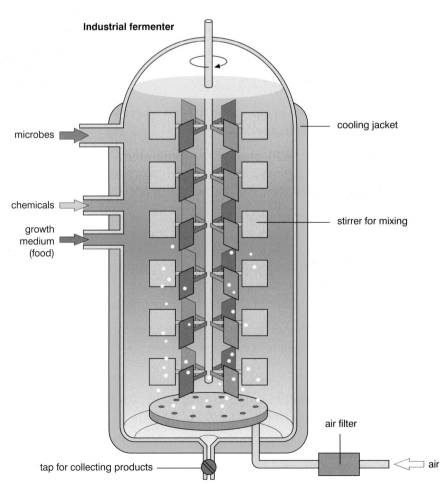

- microbes
- chemicals
- growth medium (food)
- cooling jacket
- stirrer for mixing
- air filter
- air
- tap for collecting products

- **Heat**: the reactions inside the fermenter release heat. Water passes through a cooling jacket to make sure the temperature does not rise too high. If the temperature gets too high the enzymes would be denatured.

- **Oxygen**: sterile air is pumped in at the bottom of the fermenter. Stirrers rotate the inside of the fermenter to make sure oxygen is distributed through the mixture so that aerobic respiration can take place.

- **Sterile conditions**: pumping super-heated steam through the fermenter before using it destroys any unwanted microbes.

- After fermentation, the products are taken from a tap at the bottom of the fermenter.

- The products can then be purified, separated, packaged and marketed.

QUESTION SPOTTER

▶ Questions about enzymes often ask for the factors that affect the rate of enzyme activity.

? CHECK YOURSELF QUESTIONS

Q1 Jayne is making yoghurt. The process involves three stages:
  A milk is pasteurised
  B the bacteria *Lactobacillus* is added to milk and cultured
  C lactic acid is released into the milk.
  a Explain why the milk is pasteurised.
  b Suggest why only a small amount of *Lactobacillus* is added to the milk.
  Jayne heated the mixture to 80°C.
  c Explain why the mixture did not turn into yoghurt.

Q2 The fungus *Penicillium* is used to make the antibiotic penicillin in a fermenter.
  a Explain what would happen to penicillin production if the fermenter was kept at:
  i 0°C
  ii 65°C.
  b Describe the other conditions necessary for the production of penicillin.

Q3 Industrial fermenters are filled with nutrients and a small amount of a microbe, such as bacteria. This process is used to make useful chemicals on a large scale.
  a Suggest why only a small amount of microbe needs to be used.
  b Explain why microbes such as bacteria are suitable for use in industrial fermenters.
  c Suggest **two** useful chemicals that can be produced using industrial fermenters.

Answers are on page 148.

Farming methods

What is intensive farming?

- **Intensive farming** means trying to produce as much food as possible from the land available.

INTENSIVE ANIMAL FARMING

- Cattle, pigs, sheep and hens can be kept in special units. These animals grow more quickly (or produce more eggs) because they:

  - are kept warm

  - have a minimum amount of space to move around in, so cannot use much energy

  - are safe from predators

  - have special high protein diets with many additives

  - are given antibiotics to reduce the spread of disease.

Battery hens are an example of intensive farming.

- The intensive farming of animals has reduced the cost of producing meat and eggs.

- Fish farming is also a form of intensive farming. Large cages or special pools are used to keep salmon and trout in restricted areas.

- To make sure they get a **maximum** crop yield from the plants they grow, farmers use:

  - selectively bred crops

  - artificial fertilisers

  - pesticides and herbicides

  - large fields without hedges

  - controlled conditions.

- Controlled environmental conditions for growing are used inside greenhouses. The controlled conditions increase the rate of photosynthesis. The conditions that are controlled include:

  - temperature

  - light

  - carbon dioxide

  - water.

- These factors can be controlled by a computer.

- You can find out more about farming on pages 91–93.

- Most of the tomatoes, peppers and cucumbers eaten in the UK are grown without soil. This method of growing is called **hydroponics**.

- A common method of hydroponics is the **nutrient film technique**. Plants are supported in sterile rockwool or sand. A solution can then be circulated around the roots of the plants. The circulating solution has:

  - different types and amounts of mineral salts, according to the different crops

  - oxygen bubbled into it

  - different pH conditions, according to the different crops.

QUESTION SPOTTER

▸ You will be expected to relate your understanding of photosynthesis to increasing crop yield.

These lettuces are being grown by hydroponics.

▸ It is always best to support
your answer with examples.
Be prepared to help your
explanation of biological
control by describing one
example.

▸ Organic farmers are
concerned about chemicals
entering our food chain.
Organically grown products
are becoming more popular.

▸ You need to understand why
organic foods are more
expensive than intensively
farmed foods.

⌂ What is organic farming?

- Organic farmers do not use artificial fertilisers, pesticides or herbicides on their land. Organic farmers allow animals to range freely on their farmland and supplement their diets only with natural animal feed.

- Organic farmers help to reduce both pests and weeds by growing crops with a **crop rotation system**.

- **Biological control** is used to control pests. Examples of biological control include:

  • ladybirds used to control aphids on greenhouse plants

  • lacewing insects used to control greenhouse pests

  • selective breeding of pest-resistant crops

  • some bacteria and fungi used to control pests such as canker worm.

- You can find out more about biological control on page 92.

- **Weeds** are plants growing where they are not wanted.

- Weeds are a problem for organic farmers. Removing weeds by hand takes a lot of time. Machines are being developed to recognise crop plants. This allows the machine to remove any weeds but not the crops.

- More people and more land per animal are required for organic farming methods. This causes the price of organically produced food to be more expensive.

Free-range hens.

? CHECK YOURSELF QUESTIONS

Q1 Suggest the conditions that need to be controlled to encourage fish to grow as fast as possible.

Q2 Describe the conditions that can be controlled in a greenhouse that cannot be controlled in a field.

Q3 Many people have strong views about intensive farming of animals. Suggest advantages and disadvantages of intensive farming for the consumer.

Q4 Explain why weed control is more of a problem for farmers using organic methods than for farmers using intensive methods.

Q5 Explain why food produced by organic farming methods is more expensive than food produced by intensive farming methods.

Q6 **a** Explain what is meant by biological control.
b Describe one method of biological control.

Answers are on page 149.

■ Human influences on the ■ environment

⊏⊐ Why is the effect of humans on the environment increasing?

- An **increasing human population** means that more **resources** are needed and used, such as land, raw materials for industry, sources of energy and food.

- **Technological advances** and an overall increase in the standard of living also mean that more resources are used.

- Following on from this there is an increase in waste production: **pollution**.

- These problems are caused particularly by the developed countries of the world.

⊏⊐ Consequences of misuse of the environment

- The use of some resources, such as minerals and fossil fuels, cannot continue forever: there is only a limited amount of them (they are **finite**). Once they have been used they cannot be replaced. In the future we will have to recycle these materials or find replacements that are renewable.

- Pollution caused by putting waste into the environment is causing changes that may be irreversible.

WAYS OF TACKLING THESE PROBLEMS

- The amount of **raw materials** we use should be **reduced**, for example by using less packaging on products.

- More materials should be **recycled**, so reducing the need for, and energy cost of, extracting more raw materials.

- We should be more **energy efficient**, for example by reducing heat losses from buildings or turning off lights when they are not needed.

- We should use **renewable energy sources** that will not run out and that cause little pollution, for example solar, wind and wave power.

- We should use **sustainable** or renewable **raw materials** (such as wood) that can be produced again.

⊏⊐ Farming and use of agrochemicals

- **Insecticides** are chemicals that kill insects that damage crops. These chemicals can cause unwanted effects by killing animals that are not pests.

☆ IDEAS AND EVIDENCE

▶ The increasing human population is encouraging farmers to produce as much food as possible from a reduced amount of land. This is called intensive farming. The amount of food produced is called productivity.

▶ There is an increasing interest in organic farming. This farming method does not rely on chemicals for fertilisers or pest control. You need to be able to explain why organic food is becoming more popular.

☆ IDEAS AND EVIDENCE

▶ You need to be aware of how science is used in our environment.

▶ Environment agencies try to make people aware of the need for conservation of the world's resources. Their policies emphasise the '3 Rs': Reduce, Re-use, Recycle.

▶ Everyone can take an active part by following the 3 Rs. You should be able to apply this policy to examples given in your exams.

- Some insecticides can enter the food chain. The animals that ingest them cannot break them down or excrete them, so they remain in the animals' bodies. Such substances are described as being **persistent**. Predators can contain much higher levels of the substance than animals below them in a food chain.

- Use of the persistent insecticide **DDT** is the reason why many birds of prey, such as sparrowhawks and peregrine falcons, suffered in Britain in the 1950s and 1960s. DDT is now banned in many countries. Most modern insecticides are not persistent and break down naturally in the environment after a short while.

DDT affected many food chains. In this one, the figures give the relative concentration of DDT at each stage.

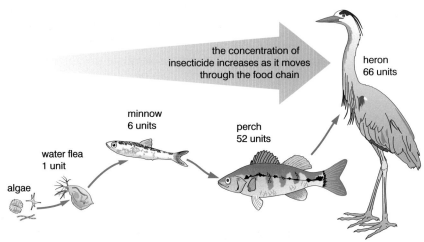

the concentration of insecticide increases as it moves through the food chain

heron
66 units

minnow
6 units

perch
52 units

water flea
1 unit

algae

- **Biological control**, introducing another organism that will kill the pests, is an alternative to using chemical insecticides. Biological control does not cause pollution and pests cannot become resistant to it (as they can to chemical insecticides). However, biological control can have unwanted effects, for instance the introduced predators might also attack harmless or even useful animals.

- **Fungicides** are chemicals that kill the fungi responsible for many plant diseases. Insecticides and fungicides together are often known as pesticides.

Ladybirds can be encouraged because they eat pest species.

- **Herbicides** (weedkillers) are chemicals that kill plants (weeds) that would otherwise grow among the crop plants and compete with them for light, water and minerals. These weeds might also be the home of other pests. However, the weeds could be used by other animals. Also, by increasing the crop, any pests that normally feed on the crop will have more food and may rapidly increase in numbers, which could cause a bigger problem than the weeds.

- **Fertilisers** are added to improve the growth of crops. **Organic** (natural) fertilisers include compost and manure. **Inorganic** (artificial) fertilisers are mined or manufactured to contain the necessary minerals. Artificial fertilisers come in powdered or liquid form so they are easy to spread, but they can cause pollution.

EUTROPHICATION

- When it rains, soluble fertiliser dissolves in the water and is carried away into lakes and rivers: it is **leached** from the soil.

- The fertiliser causes excessive growth of plants in rivers and lakes, especially **algae**. Plants growing over the surface of the water makes the water murky and blocks much of the light, and plants under the surface die and decay. The bacteria that cause the decay use up the oxygen in the water, so fish and other water animals die.

- This whole process is called **eutrophication**.

- Eutrophication can be reduced by using **less fertiliser**. This may actually help the farmer because there is a point beyond which adding more fertiliser will not help, and may even harm, crop growth.

This river is blocked with algae because of excess nitrates washed in from nearby fields.

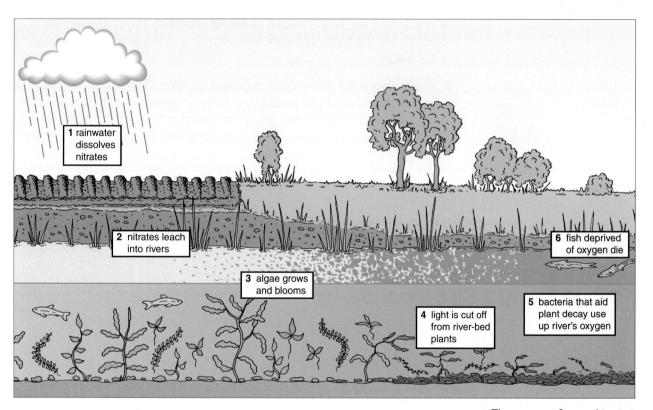

1 rainwater dissolves nitrates

2 nitrates leach into rivers

3 algae grows and blooms

4 light is cut off from river-bed plants

5 bacteria that aid plant decay use up river's oxygen

6 fish deprived of oxygen die

The process of eutrophication.

OTHER WAYS OF MAINTAINING SOIL FERTILITY

- **Organic fertilisers** should be used, because they release minerals into the soil at a slower rate than inorganic fertilisers. Organic fertilisers also help improve soil structure, for example making it easier to plough.

- A system of **crop rotation** should be used that includes growing crops that contain nitrogen-fixing bacteria in their root nodules. This restores soil fertility when the crops or their roots are ploughed back into the soil.

☐ The greenhouse effect

■ Carbon dioxide, methane and CFCs are known as **greenhouse gases**. The levels of these gases in the atmosphere are increasing due to the burning of fossil fuels, pollution from farm animals and the use of CFCs in aerosols and refrigerators.

■ Short-wave radiation from the Sun warms the ground and the warm Earth gives off heat as long-wave radiation. Much of this radiation is stopped from escaping from the Earth by the greenhouse gases. This is known as the **greenhouse effect**.

■ The greenhouse effect is responsible for keeping the Earth warmer than it otherwise would be. The greenhouse effect is normal, and important for life on Earth. However, it is thought that increasing levels of greenhouse gases are trapping more heat than is normal, and the Earth is warming up. This is known as **global warming**. If global warming continues the Earth's climate may change and sea levels rise as polar ice melts.

■ The temperature of the Earth is gradually increasing, but we do not know for certain if the greenhouse effect is responsible. It may be that the observed rise in recent global temperatures is part of a natural cycle: there have been ice ages and intermediate warm periods before. Many people are concerned that it is not part of a cycle and say we should act now to reduce emissions of greenhouse gases.

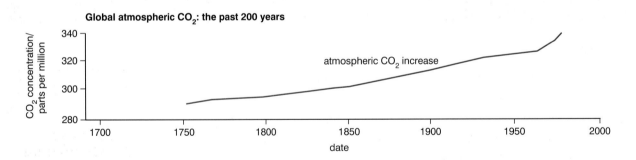

Global atmospheric CO_2: the past 200 years

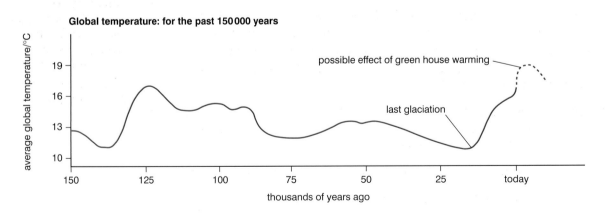

Global temperature: for the past 150 000 years

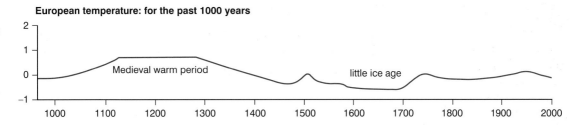

European temperature: for the past 1000 years

Medieval warm period

little ice age

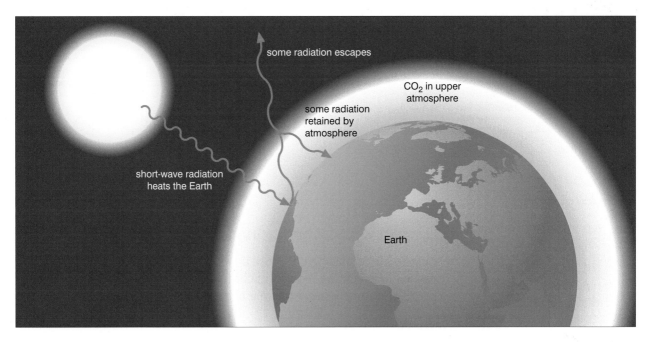

some radiation escapes

CO_2 in upper atmosphere

some radiation retained by atmosphere

short-wave radiation heats the Earth

Earth

 Acid rain

■ Burning fossil fuels gives off many gases, including **sulphur dioxide** and various **nitrogen oxides**.

■ Sulphur dioxide combines with water to form **sulphuric acid**. Nitrogen oxide combines with water to form **nitric acid**. These substances can make the rain acidic (**acid rain**).

■ Acid rain **harms plants** that take in the acidic water **and animals** that live in affected rivers and lakes. Acid rain washes ions, such as calcium and magnesium, out of the soil, **depleting the minerals available to plants**. It also washes **aluminium**, which is poisonous to fish, out of the soil and into rivers and lakes.

■ Reducing emissions of the gases causing acid rain is expensive, and part of the problem is that the acid rain usually falls a long way from the places where the gases were given off.

■ Fitting **catalytic converters** stops the emission of these gases from cars.

■ Sulphur impurities caused from burning fuels can be reduced by:

  • treating fuels before burning

  • modifying chimneys of power stations by adding limestone filters, which neutralise the acid.

QUESTION SPOTTER

▸ Your exam paper will contain data-handling style questions. You will be expected to interpret graphs and tables of data. It is common to find data-handling style questions being used to examine this section of the course.

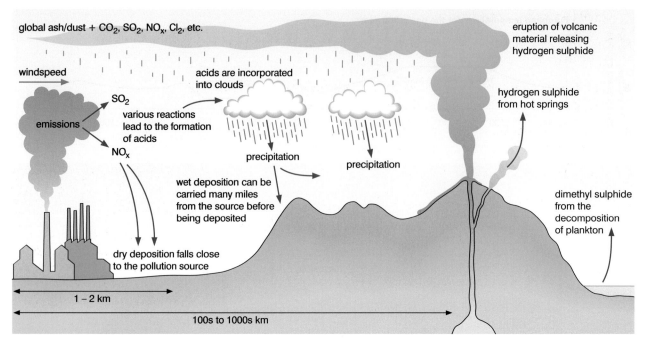

global ash/dust + CO_2, SO_2, NO_x, Cl_2, etc.

eruption of volcanic material releasing hydrogen sulphide

windspeed

acids are incorporated into clouds

hydrogen sulphide from hot springs

emissions

SO_2

various reactions lead to the formation of acids

NO_x

precipitation

precipitation

wet deposition can be carried many miles from the source before being deposited

dimethyl sulphide from the decomposition of plankton

dry deposition falls close to the pollution source

1 – 2 km

100s to 1000s km

The problem of acid rain.

⬚ Depletion of the ozone layer

- High in the atmosphere there is a layer of a gas called **ozone**, a form of oxygen in which the molecules contain three oxygen atoms (O_3). Ozone in the atmosphere is continually being broken down and reformed:

ozone formation	ozone breakdown
$O_2 \xrightarrow{\text{UV light}} O + O$ $O_2 + O \rightarrow O_3$	$O_3 \rightarrow O_2 + O$ $O_3 + O \rightarrow 2O_2$

- The ozone layer reduces the amount of **ultraviolet radiation (UV)** that reaches the Earth's surface. This is important because UV radiation can cause skin cancer.

- Normally the rates at which the ozone forms and breaks down are about the same. **CFCs**, which are gases used as aerosol propellants and in refrigerator cooling systems, **speed up** the rate of ozone breakdown without affecting how quickly it reforms. This means that there is **less ozone** in the atmosphere and that **more UV radiation** will be reaching the Earth's surface. This could cause an increase in skin cancer in people who are exposed to a lot of sunlight.

- We can reduce the damage to the ozone layer by finding alternatives to CFCs, but many of these seem to contribute to other problems, such as the greenhouse effect.

- Even if all CFC emissions stopped today it will still take about a hundred years for all the CFCs presently in the atmosphere to break down.

Household waste

- People produce a lot of waste, including sewage and rubbish they simply throw away.

- **Sewage** has to be treated to remove disease organisms and the nutrients that cause eutrophication, before it can be discharged into the sea.

- Some **household rubbish** is burnt, causing acid gas pollution and acid rain. Rubbish tips create their own problems:

  - they are ugly and can smell

  - they can encourage rats and other pests

  - methane gas produced by rotting material may build up in tips that are covered with soil and this gas is explosive

  - covered-over tips cannot be used for building on because the ground settles.

- We can reduce the amount of material in our dustbins by **recycling** or **reusing** materials and **not buying highly packaged** materials.

QUESTION SPOTTER

▸ Questions about environmental damage may use examples that you have not studied in detail. You will be given enough information to be able to answer them.

Habitat destruction

- Many natural habitats are being **destroyed** to create land for farming or building on. The rainforests are being cut down for their wood and to create farming or grazing land. Many species of animals and plants are only found in wetland areas, which are being drained to 'reclaim' the land.

- Removing plants exposes the soil to rain, which **washes the soil away**, blocking rivers and causing flooding.

- Habitat destruction can also **alter the climate** because less water is transpired into the atmosphere.

- Destruction of habitats **reduces species diversity**.

IDEAS AND EVIDENCE

▸ Rainforests are being destroyed to make room for arable farmers and cattle ranchers. This means fewer trees are left to remove the carbon dioxide from the atmosphere and less oxygen is being released into the atmosphere.

▸ You need to be able to discuss whether people should be allowed to destroy areas of rainforest to earn their living. Be prepared to support your views with scientific facts.

An example of a habitat destroyed by human actions.

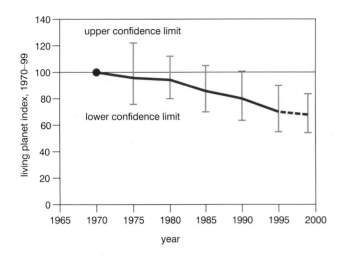

The species diversity index

- The **species diversity index** is a useful way to **summarise data** that are being used to indicate the diversity of a particular habitat. A habitat with good conditions for many species to feed and reproduce in, such as a tropical rainforest, is expected to have a high species diversity index. A habitat with more harsh conditions, such as a desert, is expected to have a lower species diversity index because fewer species can survive there.

An example of a species diversity index: WWF Living Planet Report 2000.

QUESTION SPOTTER

▸ You will not be expected to calculate a species diversity index, but you need to be able to explain what it is used to indicate.

Endangered species

- Many species of animals and plants are in danger of **extinction** because of habitat destruction and hunting.

- Endangered species can be helped by:
  - protecting areas where they live
  - legal protection, such as laws against hunting
  - educating people so that they are aware of species that are endangered and know how to protect them.
  - eggs, sperm and embryos of endangered species being **frozen** and stored in **gene banks**
  - **captive breeding programmes**, which are being carried out in many zoos to breed rare animals in captivity then release the offspring into natural habitat
  - creating **artificial ecosystems**, such as safari parks and zoos.

Conservation

- **Conservation** means to protect different species of plants and animals and their environments, to keep their populations healthy and reproducing.

- We need to protect ecosystems so that many different species can survive. This is usually part of a **conservation programme**. Special conservation programmes are often needed to help endangered species survive.

- The survival of the Hawaiian goose was helped with a conservation programme. In 1950 this bird was almost extinct in Hawaii because of human activities. A few breeding pairs were brought to the Wildfowl Trust Centre at Slimbridge in Gloucestershire, UK. The geese bred successfully in this semi-natural ecosystem. In the early 1960s the geese had increased in numbers sufficiently for some to be returned Hawaii. Since some geese were released in Hawaii, their numbers have gradually increased.

REASONS FOR CONSERVATION

- We may need to conserve ecosystems because they contain **a human food supply**, for example sea fish. The main way of catching fish in the sea is by trawling. Laws exist about the size of net, times of year and fish quotas for trawling so that we do not remove the fish faster than they can be replaced by the natural process of reproduction. At the moment we are killing and using more cod than are being produced, so further conservation is needed.

- We also need to make sure that there is **no damage being done to the food chains** that fish, for example, depend on.

- Another important reason for conservation is our use of **plants for medical purposes**. There are many plants in the rainforests that have yet to be identified. Many plants of the rainforests have healing powers. If we are not aware of the plants we may not be aware of a cure for illnesses such as cancer.

- Another reason for conservation is to maintain **cultural resources**. Many people enjoy nature reserves. Conservation is important in ensuring the aesthetic qualities of an area are maintained.

QUESTION SPOTTER

▸ When answering a question about conservation, be prepared to support your answer with an example, such as that of the Hawaiian goose.

IDEAS AND EVIDENCE

▸ You need to be aware of and understand the cultural and economic effects of conservation. Laws are passed, for example, to limit the number of animals that can be hunted, or hunting is banned. This has an effect on many people in developing countries where hunting is one of the very few ways they have of feeding themselves or earning money.

▸ Financial and educational programmes may need to be introduced to support these laws. If people are made aware of the problems there may be an improvement. However, educational programmes cost a lot of money.

Sustainable development

- One way of reducing the impact of human activity on the environment is to **replace** the materials or crops we use.

- For example, quotas have been placed on fishing catches so that fish stocks do not drop so far the fish disappear altogether.

- Another example is to replant woodland after trees have been cut down.

CHECK YOURSELF QUESTIONS

Q1 Why are some pesticides described as 'persistent'?

Q2 What is the difference between the greenhouse effect and global warming?

Q3 Acid rain can be reduced by passing waste gases through lime to remove sulphur dioxide. Why do all factories not automatically do this?

Q4 a Explain the difference between natural and artificial ecosystems.
b Describe a situation where it could be necessary to create an artificial ecosystem.

Q5 a Describe what the species diversity index is used for.
b Describe a habitat that would have:
  i   a low species diversity index
  ii a high species diversity index.

Q6 a Explain why laws are passed to help conserve species of plants and animals.
b Describe other issues that need to be taken into account when conservation laws are passed.

Answers are on page 149.

UNIT 9: GENETICS AND EVOLUTION

REVISION SESSION 1  Variation

Types of variation

- No two people are the same. Similarly, no two oak trees will be exactly the same in every way, they will have different heights, different trunk widths and different numbers of leaves.

- There are two types of variation:

  1 **discontinuous variation** (sometimes called discrete variation) – a characteristic can have one of a certain number of specific alternatives, for example gender, where you are either male or female, and blood groups, where you are either A, B, AB or O

  2 **continuous variation** – a characteristic can have any value in a range, for example body weight and length of hair.

Causes of variation

- **Environmental causes** include your diet, the climate you live in, accidents, your surroundings, the way you have been brought up and your lifestyle. They all influence your characteristics.

- **Genetic causes** are the characteristics controlled by your genes. Genes are inherited from your parents. Examples of characteristics in humans influenced purely by genes are eye colour and gender.

- Many characteristics are influenced by environment **and** genes. For example people in your family might tend to be tall, but unless you eat correctly when you are growing you will not become tall, even though genetically you have the tendency to be tall. Other examples are more controversial, such as human intelligence, where it is unclear whether the environment or genes is more influential.

Genes

- Genes are chemical **instructions** that direct the processes going on inside cells. They affect the way cells grow and work and so can affect features of your body such as the shape of your face and the colour of your eyes. All the information needed to make a fertilised egg grow into an adult is contained in its genes.

- Inside virtually every cell in the body is a **nucleus**, which contains long threads called **chromosomes**. These threads are usually spread throughout the nucleus, but when the cell splits they gather into bundles that can be seen through a microscope. The chromosomes are made of a chemical called **deoxyribonucleic acid (DNA)**.

> ### IDEAS AND EVIDENCE
>
> ▸ Francis Galton thought that all intelligence was inherited. He believed that environmental factors, such as parents and teachers, had nothing to do with how intelligent people are.
>
> ▸ You need to be able to give your own thoughts on this issue. Make sure you can provide scientific reasons to support your thoughts.

> ### A* EXTRA
>
> ▸ DNA is made of very long strings of four different chemical bases.
> ▸ Each group of three bases codes for a particular amino acid, so the sequence of bases in a gene codes for the sequence of amino acids in a protein.

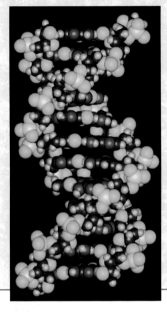

▸ The 'double-helix' structure of DNA was discovered by James Watson and Francis Crick working in Cambridge in 1953. Rosalind Franklin, working in London, contributed important work on the structure of DNA.

DNA is found in the nucleus of cells.

■ DNA is a very long molecule that contains a series of chemicals called **bases**. There are four different types of base in DNA: **thymine (T)**, **adenine (A)**, **guanine (G)** and **cytosine (C)**. C and G always match together and A and T always match together. The order in which the bases occur is a **genetic code**. The code spells out instructions that control how the cell works. Each length of DNA that spells out a different instruction is known as a **gene**. Each chromosome contains thousands of genes.

■ All the genes in a particular organism are known as its **genome**.

■ Genes tell cells how to make different **proteins**. Many proteins are enzymes that control the chemical processes inside cells.

■ Only some of the full set of genes are used in any one cell.

■ Most of our features that are controlled genetically are affected by several genes.

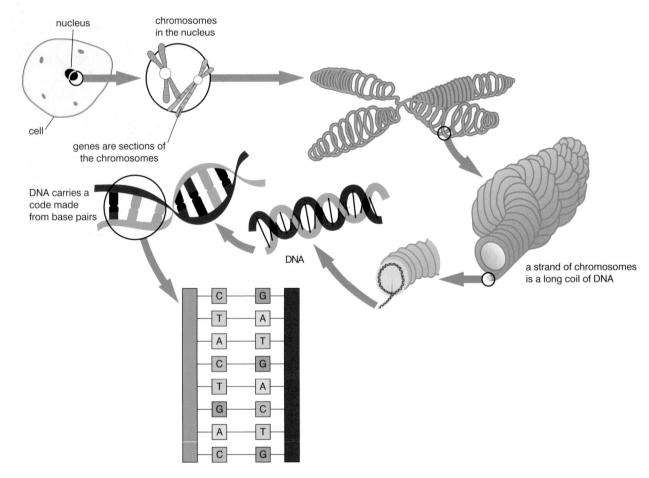

nucleus

chromosomes in the nucleus

cell

genes are sections of the chromosomes

DNA carries a code made from base pairs

DNA

a strand of chromosomes is a long coil of DNA

C	G
T	A
A	T
C	G
T	A
G	C
A	T
C	G

Mutations

- Sometimes genes can be altered so that their message becomes meaningless or a different instruction. This is known as a **mutation**. Mutations can be caused by:

  - radiation

  - certain chemicals, called **mutagens**

  - spontaneous changes.

- Most mutations are harmful. A few mutations are beneficial to the plant or animal and can increase its population by natural selection. (There is more about natural selection later in this unit, in Revision Session 5.)

- A single mutated gene causes the inherited disease **cystic fibrosis**.

- An extra chromosome causes **Down's syndrome**. A person with Down's syndrome has 47 chromosomes instead of the usual 46.

- Mutations that occur in **sex cells** can be passed on to offspring, who may develop abnormally or may die at an early stage of development.

- If mutations occur in **body cells** they may multiply uncontrollably. This is **cancer**.

Cell division

- Cells grow by splitting in two. This is called **cell division**, and is part of normal body growth. It is also the way that single-celled organisms reproduce and is the only type of cell division involved during **asexual reproduction** (reproduction that does not involve sex cells).

- Before a cell splits, its chromosomes **duplicate** themselves. The new cells formed, sometimes called the **daughter cells**, contain chromosomes identical with the original cell. This type of cell division, in which the new cells are genetically identical to the original, is known as **mitosis**. Cells or organisms that are genetically identical to each other are known as **clones**.

- Mitosis takes place in all the normal body cells.

QUESTION SPOTTER

▸ You are likely to be asked about the difference between genes and chromosomes and where they are located in a cell.

A spider plant forms new plants at the end of stalks ('runners'). These new plants eventually grow independently of the parent plant. This is an example of asexual reproduction.

Stage 1. The chromosomes get fatter and become visible.

Stage 2. Each chromosome makes an exact copy of itself.

Stage 3. Remember that each chromosome has its own 'partner'. They carry similar 'blueprints' – one from the original father and one from the mother. These partner chromosomes get together and line up across the middle of the cell.

Stage 4. The colour chromosomes part and move to opposite halves of the cell.

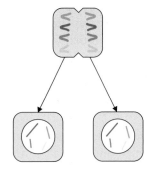

Stage 5. The cell splits in two. These new cells have the same number of chromosomes as the cell at the start of the process.

- There is another type of reproduction in which the new individuals have been formed by two special cells, called **sex cells** or **gametes**, usually one from each parent. This type of reproduction is called **sexual reproduction**.

- During sexual reproduction, a different type of cell division called **meiosis** makes sure that the gametes only have half the number of chromosomes that body cells have. In humans the gametes are **sperm** cells and **egg** cells. In flowering plants the gametes are the **pollen** cells and the female 'egg' cells (**ovules**).

A* EXTRA

▸ In meiosis, cells divide twice, forming four gametes (sex cells), each with a single set of chromosomes. The new cells are not genetically identical.

▸ In mitosis, cells divide once, producing two cells, each with a full set of chromosomes. The new cells are genetically identical to each other and to the original cell.

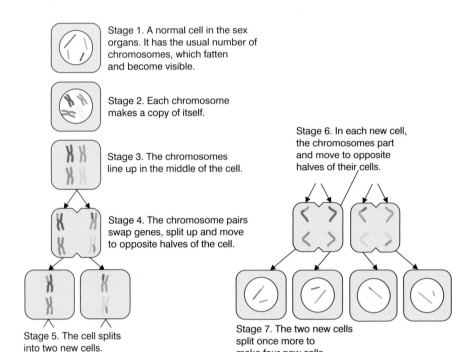

Stage 1. A normal cell in the sex organs. It has the usual number of chromosomes, which fatten and become visible.

Stage 2. Each chromosome makes a copy of itself.

Stage 3. The chromosomes line up in the middle of the cell.

Stage 4. The chromosome pairs swap genes, split up and move to opposite halves of the cell.

Stage 5. The cell splits into two new cells.

Stage 6. In each new cell, the chromosomes part and move to opposite halves of their cells.

Stage 7. The two new cells split once more to make four new cells. Each has half the usual number of chromosomes.

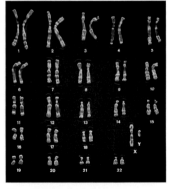

The 23 pairs of chromosomes in a human cell, including the X and tiny Y.

- The gametes contain half the normal number of chromosomes so that when they join together (at **fertilisation**) the normal number is restored. For example, human cells normally contain 23 pairs of chromosomes: 46 in total (**diploid**). Human egg and sperm cells contain only 23 chromosomes (**haploid**), so when they join together the new cell formed has the normal 46. (The number of chromosomes depends on an organism's species.)

- The offspring produced by sexual reproduction are genetically **different** from their parents.

- Every gamete formed has one of each pair of chromosomes. However, which one of each chromosome pair ends up in a particular gamete is purely random. An individual human can produce gametes with billions of different combinations of chromosomes.

- Added to this variation is the fact that when chromosome pairs are lying next to each other they swap lengths of DNA (**crossing over**), altering the combination of genes on a chromosome. It is therefore not surprising that, even with the same parents, we can look very different from our brothers and sisters.

Genetic fingerprinting

- **Genetic fingerprinting** can now be used to identify a person from a sample of body tissue. This is done by:
  - extracting DNA from cells
  - cutting up the DNA using enzymes
  - separating the DNA fragments with a process similar to chromatography
  - labelling the DNA fragments using a radioactive substance
  - X-raying the sample
  - comparing the patterns formed from different DNA samples.

IDEAS AND EVIDENCE

- Allan Wilson of the University of California studied the DNA of 147 people. He identified similarities in their DNA that suggested they were all descended from one woman.
- Further studies have supported Wilson's findings. These studies show that we are all descended from a female who lived some 200 000 years ago.

CHECK YOURSELF QUESTIONS

Q1 Which of the following are examples of (a) continuous and (b) discontinuous variation?

A Hair colour.
B Blood group.
C Foot size.
D Gender.
E Hair length.

Q2 Bob and Dave are identical twins. Dave emigrates to Australia, where he works outside on building sites. Bob stays in England and works in an office. Suggest how they might look (a) similar and (b) different. Explain your answers.

Q3 Put the following in order of size, starting with the smallest: cell nucleus, chromosome, gene, cell, chemical base.

Answers are on page 150.

REVISION SESSION 2

■ Genes and proteins ■

QUESTION SPOTTER

▶ If you are doing the Edexcel Higher Tier exam you need to be able to refer to the mRNA inside the nucleus and the tRNA at the ribosomes in the cytoplasm.

☐ How are proteins made?

■ **DNA** is divided into sections called **genes** (see page 102). One gene contains the code for one protein. The DNA does not make new proteins itself, this is done by **RNA**. RNA **synthesises** (makes) proteins.

1 An RNA **template** is made of the gene.

2 The RNA template contains the codes for a protein. The RNA moves out of the nucleus into the cytoplasm.

3 Along the RNA template, **three bases** provide the code for **one amino acid**. A **chain** of amino acids is formed to produce the new protein molecule.

Proteins are synthesised by RNA within a cell.

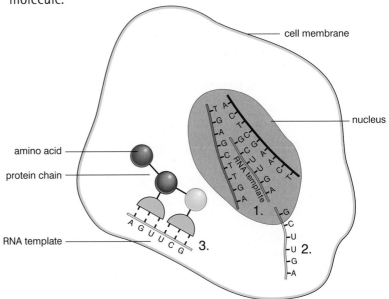

☐ Recombinant DNA

■ All living things use the same **four base codes** for genes. This means the language of DNA is universal to every living thing. It is possible to take a gene from one living organism and place it into the DNA of a different organism. The instructions from the gene will be understood and the protein will be made from the gene (see genetic engineering on pages 113–114).

CHECK YOURSELF QUESTIONS

Q1 Describe the stages of protein synthesis.

Q2 Explain the term 'one gene – one enzyme'.

Q3 Explain why DNA can be taken from one organism and used in a different organism.

Q4 There are only four base codes for DNA. Explain why it is possible for only four base codes to be used to produce more than 20 different amino acids.

Answers are on page 151.

▰ Inheritance ▰

⌂ How features are passed on

- Some features, such as the colour of your eyes, are passed on (**inherited**) from your parents, but other features may not be passed on. Sometimes features appear to miss a generation, for instance you and your grandmother might both have ginger hair, but neither of your parents do.

DOMINANT AND RECESSIVE ALLELES

- Leopards occasionally have a cub that has completely black fur instead of the usual spotted pattern. It is known as a black panther but is still the same species as the ordinary leopard.

- Just as in humans, leopard chromosomes occur in **pairs**. One pair carries a gene for fur colour. There are two copies of the gene in a normal body cell (one on each chromosome). Both copies of the gene may be identical but sometimes they are different, one being for a spotted coat and the other for a black coat. Different versions of a gene are called **alleles**.

- Leopard cubs receive half their genes from each parent. Eggs and sperm cells only contain half the normal number of chromosomes as normal body cells. This means that egg and sperm cells contain only one of each pair of alleles. When an egg and sperm join together at fertilisation the new cell formed, the **zygote**, which will develop into the new individual, now has two alleles of each gene.

- **Different combinations** of alleles will produce different fur colour:

  spotted coat allele + spotted coat allele = spotted coat

  spotted coat allele + black coat allele = spotted coat

  black coat allele + black coat allele = black coat

- The black coat only appears when **both** of the alleles for the black coat are present. As long as there is at least one allele for a spotted coat, the coat will be spotted because the allele for a spotted coat overrides the allele for a black coat. It is the **dominant allele**. Alleles like the one for the black coat are described as **recessive**.

- Characteristics caused by two identical alleles are called **homozygous**.

- Characteristics caused by two different alleles are called **heterozygous**.

MONOHYBRID CROSSES

- An individual's combination of genes is his or her **genotype**. An individual's combination of physical features is his or her **phenotype**. Your genotype influences your phenotype.

- We can show the influence of the genotype in a **genetic diagram**. In a genetic diagram we use a **capital letter** for the **dominant** allele and a **lower case letter** for the **recessive** allele.

> ✬✬ **IDEAS AND EVIDENCE**
>
> ▸ You need to be aware of major scientific developments.
>
> ▸ The Human Genome Project was an international project lead by John Sulston of Cambridge University. Scientists all over the world took part in the project to map the genes of human DNA. The project was completed in 2001.
>
> ▸ You will need to consider how this project could be useful to various people, for example whether life insurance companies should have access to the data to see whether someone has inherited a genetic disorder. You also need to consider whether this information should be made available to everyone.

▸ In your exam you could be asked to use genetic diagrams to show the results of a cross.

▸ Unless you are asked otherwise, make sure you show clearly all the possible gametes (sex cells) and offspring, and how they have been produced.

■ Using the example of the leopards, the letter S stands for the dominant allele for a spotted coat and letter s stands for the recessive allele for the black coat. Two spotted parents who have a black cub must each be carrying an S and an s. The genetic diagram below shows the different offspring that may be born.

	mother		father
parents:	**Ss**		**Ss**
	(spotted)		(spotted)

gametes: **S** or **s**   **S** or **s**

first generation, also known as F_1:

SS   or   **Ss**   or   **Ss**   or   **ss**
(spotted)   (spotted)   (spotted)   (black)

■ If the first generation, F_1, was crossed again, the offspring of this second generation would be referred to as F_2.

■ Because we are looking at only one characteristic (fur colour), this is an example of a **monohybrid cross**. 'Mono' means one, and a 'hybrid' is produced when two different types breed or cross.

■ Another type of genetic diagram is known as a **Punnett square**.

■ The example above can be shown in a Punnett square:

		male gametes	
		S	**s**
female gametes	**S**	SS (spotted)	Ss (spotted)
	s	Ss (spotted)	ss (black)

A* EXTRA

▸ A genetic diagram (such as a Punnett square) shows the probabilities of the different possible outcomes of a cross. Actual results may well differ.

IDEAS AND EVIDENCE

▸ How genes are inherited was first worked out in the nineteenth century by an Austrian monk, Gregor Mendel, who did lots of breeding experiments using pea plants.

▸ Mendel presented the results of his work to the Natural History Society in 1865. Unfortunately no one could offer an explanation for the results. It was not until powerful microscopes had been developed and chromosomes had been discovered that Mendel's work could be explained.

■ In a diagram of a monohybrid cross, when two heterozygous parents are crossed, the offspring that have the feature controlled by the dominant allele and the offspring that have the feature controlled by the recessive allele appear in a **3:1 ratio**. This is because of the random way that gametes combine. The 3:1 ratio refers to the **probabilities** of particular combinations of alleles, so, for example, there is a 1 in 4 chance of a leopard cub being black.

■ With a large number of offspring in an actual cross of two heterozygous leopard parents, you would expect something near the 3:1 ratio of spotted to black cubs. However, because it does only refer to probabilities you should not be too surprised if, for example, in a litter of four, two of the cubs were black or none was black.

■ The genotypes of individuals can be worked out by using a **family tree**.

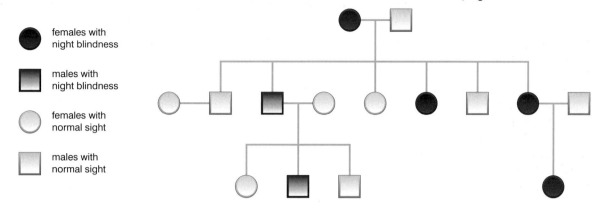

females with night blindness

males with night blindness

females with normal sight

males with normal sight

Inherited diseases

- Some diseases or disorders can be inherited. Examples include:

  - **cystic fibrosis** – in which the lungs become clogged up with mucus

  - **haemophilia** – in which blood does not clot as normal

  - **diabetes** – in which insulin is not made in the pancreas (see Unit 4)

  - **sickle–cell anaemia** – in which the red blood cells are misshapen and do not carry oxygen properly

  - **Huntington's disease** – a disorder of the nervous system.

- Most inherited diseases are caused by **faulty genes**. For example, one form of diabetes is caused by a fault in the gene carrying the instructions telling the pancreas cells how to make insulin.

- Most of these faulty alleles are **recessive**, which means that you have to have two copies of the faulty allele to show the disorder. Many people will be **carriers**, having one normal allele as well as the faulty version. The diagram below shows how two carrier parents could have a child with a disorder. The probability of these parents having a child with diabetes is a 1 in 4 chance, or 25%.

IDEAS AND EVIDENCE

- ▸ By ten weeks into a pregnancy, it is possible for doctors to predict whether a pregnant woman is carrying a baby with Down's syndrome. If the baby has Down's syndrome, the mother is offered a termination.

- ▸ The mother is then left in a moral dilemma in trying to decide whether to have the pregnancy terminated.

		male gametes	
		D	**d**
female gametes	**D**	**DD** (makes insulin)	**Dd** (makes insulin)
	d	**Dd** (makes insulin)	**dd** (has diabetes)

	male gametes	
	X	**Y**
female gametes **X**	**XX** (girl)	**XY** (boy)
female gametes **X**	**XX** (girl)	**XY** (boy)

⮂ Sex determination

- Whether a baby is a boy or a girl is decided by one pair of chromosomes called the **sex chromosomes**. There are two sex chromosomes, the **X chromosome** and the **Y chromosome**. Boys have one of each and girls have two X chromosomes.

- **Egg** cells always contain **one X chromosome** but **sperm** cells have an equal chance of containing **either** an X chromosome **or** a Y chromosome. This means that a baby has an equal chance of being either a boy or a girl. This is shown in the diagram on the left.

⮂ Sex-linked inheritance

- Some genetic disorders, such as red–green colour blindness, are more common in men than women. The recessive allele causing the disorder is found on part of the X chromosome, and there is no equivalent part on the Y chromosome, because it is smaller (see the photograph on page 104), to carry either a dominant or recessive allele. This means that males only need to have one copy of the recessive allele to show the disorder, while females would need two copies.

The British Royal family tree showing sex-linked inheritance of haemophilia.

normal female

normal male

carrier female

haemophiliac male

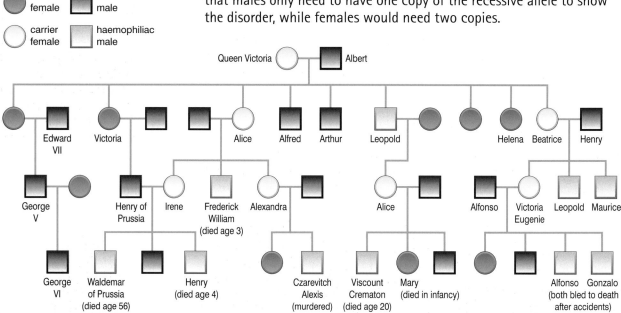

CHECK YOURSELF QUESTIONS

Q1 In humans, the gene that allows you to roll your tongue is a dominant allele. Use a genetic diagram to show how two 'roller' parents could have a child who is a 'non-roller'.

Q2 In leopards the allele for being spotted, S, is dominant to the allele for being black, s. How could you tell whether a spotted leopard was Ss or SS, given that you have black leopards that you know are ss?

Q3 John and Mary have a little girl who has cystic fibrosis. Cystic fibrosis is caused by a recessive allele. What is the probability of their next child also having cystic fibrosis? Neither John nor Mary have the condition.

Answers are on page 151.

Applications

⊡ Selective breeding

■ **Selective breeding** (also known as **artificial selection**) has been used by farmers for hundreds of years to change and improve the features of the animals and plants they grow. Examples are breeding:

- apples that are bigger and more tasty

- sheep that produce more or better quality wool

- cows that produce more milk or meat

- wheat that can be planted earlier to give an extra crop

- plants that are more resistant to disease.

■ In every case the principle is the same. You select two individuals with the features closest to those you want and breed them together. From the next generation you again select the best offspring and use these for breeding. This is repeated over many generations.

■ For example, if you wanted to produce dogs that had short tails, you would breed the male and female with the shortest tails from the dogs you had available. From their offspring you would select those with the shortest tails to use for breeding. After repeating this over many generations you would find that you almost always produced short-tailed dogs: they would be **breeding true**.

■ Simply selecting for different features such as size or shape has produced all the breeds of dogs we have today. They are all descended from wolves that were domesticated thousands of years ago.

> ⚡ **A* EXTRA**
>
> ▸ Selective breeding usually causes a reduction in variation.
> ▸ Selective breeding may also lead to an increase in harmful recessive characteristics. This is the danger of inbreeding.

The power of selection. Selective breeding has produced an enormous variety of dog breeds from wild wolf stock (right).

- To help the selective breeding process, farmers can use **artificial insemination** to breed their cattle. Sperm is collected from the best bulls and stored until required. The sperm can then be inserted into selected cows.

- Another method used to try and guarantee the desired offspring is called *in vitro* fertilisation. (*In vitro* means in glass, which here refers to the small glass dish where fertilisation takes place.) Sperm is collected from the best bulls and eggs are taken from the best cows. The sperm and eggs are placed into a small dish and fertilisation takes place. When the fertilised egg develops into a bundle of cells (**embryo**) it is transplanted into a **surrogate** mother.

⛶ Cloning

- If you have an animal or plant with features that you want, its offspring might not show those features. A simple way of producing new **plants** that are genetically identical to the original is by taking **cuttings**. A cutting is a shoot or branch that is removed from the original and planted in soil to grow on its own. Some plants do this easily, others may need rooting hormones (see Unit 6) to encourage them to grow roots.

- A more modern version of taking cuttings is **tissue culture**. This is used to grow large numbers of plants quickly as only a **tiny part** of the original is needed to grow a new plant. This method is also called **micropropagation**. Special procedures have to be followed.

1 Many small pieces are cut from the chosen plant. These pieces are called **explants**.

2 The pieces are **sterilised** by washing them in mild bleach to kill any microbes.

3 In sterile conditions the explants are transferred onto a jelly-like growth medium that contains nutrients as well as plant hormones to encourage growth.

4 The explants should develop roots and shoots and leaves.

A small cutting from a geranium has been encouraged to grow fresh roots.

These tiny sundew plants have all grown from tiny cuttings of a single plant.

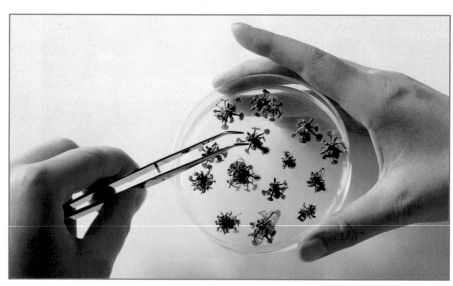

5 When the plants are large enough they can be transferred to other growth media and eventually to a normal growth medium like compost.

- Cloning can also now be used with **animals**. For example, cells from a developing **embryo** (such as from a cow) can be split apart before they become specialised and then transplanted, as identical embryos, into **host mothers**.

- Another way of cloning animals involves making ordinary **body cells** grow into new animals. Dolly the sheep was the first mammal clone grown from the cell of a fully grown organism: in Dolly's case it was a cell from the udder of a sheep.

Genetic engineering

- Selective breeding works by trying to bring together desirable combinations of genes. **Genetic engineering** allows this to be done quickly and directly. It involves taking genes from **one organism** and **inserting** them into cells of **another**. For example, genes have been inserted into crop plants to make them more resistant to attack by pests.

- The gene that carries the instructions for making the human hormone insulin has been inserted into bacteria, which then produce insulin. This is where much of the insulin to treat diabetes today comes from. The gene was originally from a human cell so the insulin is identical to that made normally in the human pancreas.

STEPS INVOLVED IN GENETIC ENGINEERING

1 **Find** the part of the DNA that contains the gene you want.

2 **Remove the gene** using special enzymes that cut the DNA either side of the gene.

3 **Insert the gene** into the cells of the organism you want to change.

- Step 3 may be done directly but it often involves using **other organisms**, such as bacteria or viruses, to act as **vectors**. This means the gene is first inserted in the vectors, which then transfer the gene to the desired host when they infect it.

- Inserting the gene into bacteria means that many copies can be made of the gene because bacteria reproduce asexually, producing new cells that are genetically identical to the original.

A* EXTRA

▶ If new genes are introduced to an animal or plant embryo in an early stage of development then, as the cells divide, the new cells will also contain copies of the new gene.

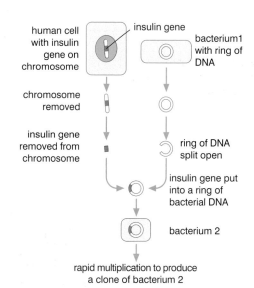

The process of genetic engineering used for insulin production.

▸ In questions about selective breeding and genetic engineering you may be asked to describe the steps involved in each process.

▸ These questions usually carry several marks, so make sure you describe all the steps.

DANGERS OF GENETIC ENGINEERING

■ Although genetic engineering promises many benefits for the future, such as curing genetic diseases and changing crops to make them disease resistant, there are possible dangers.

■ For example, the **wrong genes** could accidentally be transferred or viral infection could transfer inserted genes to other unintended organisms. Imagine the problems if, instead of crops becoming resistant to disease, weeds became immediately resistant to herbicides, or if harmful bacteria became immediately resistant to medicines like antibiotics.

IDEAS AND EVIDENCE

▸ Genetic modification of plants may allow the development of crops that could be grown in desert areas.

▸ Genetic engineering can be used to modify plants to improve production of food, for example by increasing a crop's resistance to pests or herbicides, increasing its nutritional value or extending its shelf-life.

▸ Many people think that these are examples of good reasons for developing genetically modified crops. Try to think of reasons why people disagree with the introduction of genetically modified crops into this country.

CHECK YOURSELF QUESTIONS

Q1 A farmer wants to produce sheep with finer wool. How could this be done by selective breeding?

Q2 What are the advantages and disadvantages of producing new roses by tissue culture (cloning)?

Q3 Both genetic engineering and selective breeding are ways of producing new combinations of genes in an organism. What are the advantages of genetic engineering?

Answers are on page 152.

Evolution

⌷ Evidence for evolution

■ Animal and plant species can change over long periods of time. This change is called **evolution**.

■ We cannot **prove** that evolution has happened because we cannot go back in time and watch. However, there is a lot of evidence that evolution has happened.

THE FOSSIL RECORD

■ **Fossils** show that very different animals and plants existed in the past. In some cases we can see steady **progressive changes** over time. In other cases there is no evidence of a particular organism's ancestors or descendants.

■ The likelihood of a fossil forming in the first place is very rare. Fossils may form in various ways:

  • from **hard body parts** that do not decay easily

  • from body parts that have not decayed because of the **conditions** the body was left in

  • from parts of an animal or plant that has decayed and been **replaced** by other materials

  • as **traces** of activities, such as footprints and burrows.

■ Even if fossils did form they might not be discovered or could be destroyed by erosion. If evolution happened relatively quickly there would be less chance of finding fossils of the intermediate stages. Even when fossils are found they usually only provide a record of the hard tissues such as bone.

■ A process called **carbon dating** can date fossils. This enables scientists to build up a picture of what has happened to plants and animals over a long period of time.

AFFINITIES

■ **Affinities** are similarities that exist between different species. For example, humans, dolphins and rats have similar numbers and basic arrangements of bones in their skeletons. The easiest explanation is that they are similar because they are related, in other words they have evolved from a common ancestor.

⌷ Accepting the idea of evolution

■ It was only after 1859, when **Charles Darwin** published his famous book *On the Origin of Species*, that ideas about **evolution** were widely discussed. The scientist **Alfred Russell Wallace** also came up with the same ideas as Darwin, but he did not publish them first.

 A* EXTRA

▶ The arrangement of bones in the wings of birds is useful evidence for the theory of evolution. The arrangement of bones in a bird's wing is similar to that of a human arm.

- Many people were angered by Darwin's ideas as they believed they went against the Christian Church's teachings that God had created the world and all the living things in it. Many newspaper cartoons and articles ridiculed Darwin's ideas. Some of these were based on the mistaken idea that Darwin had said that humans had evolved from apes like those present today, such as chimpanzees.

- The idea of evolution is not that the animals and plants we see around us evolved from each other, but that some have similar features because they evolved from a **common ancestor**. The diagram below illustrates how humans and apes may have evolved from a common ancestor. Humans and chimpanzees share more similarities than the other types of apes, which means that they probably share a more recent ancestor than the others.

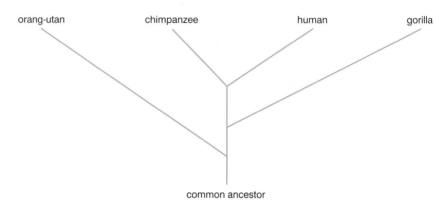

orang-utan   chimpanzee   human   gorilla

common ancestor

- Today there are still people who do not believe in the idea of evolution.

⟨⟩ Natural selection

- Darwin suggested how evolution could have happened. He called his theory **natural selection**. He based his ideas on his observations of the wildlife he found on several lengthy sea voyages. He also used his knowledge of selective breeding.

- Natural selection is the same basic idea as selective breeding, except it is 'nature' rather than the breeder that determines which animals or plants are able to reproduce and pass on their features.

The cactus is well adapted to dry desert conditions:
- *long roots reach water far below ground*
- *a swollen stem stores water*
- *leaves are small spines, which reduce water loss by having a small surface area and provide the plant with protection against animals*
- *a rounded shape reduces the surface area and therefore water loss.*

THE IDEAS OF NATURAL SELECTION

1 Animals and plants produce **more** offspring than will ever survive.

2 Population sizes remain fairly constant over time.

3 **Sexual reproduction** means there is variation. So all the offspring from a pair of parents are slightly different from each other. This means that if some individuals are slightly better **adapted** to a situation they will have a better chance of surviving than the others. For example, a zebra that could run faster than the others would be less likely to be caught by lions, or a cactus with slightly sharper spines would be less likely to be eaten. This idea is sometimes known as **survival of the fittest**.

The polar bear is well adapted to its environment:
* eyes at the front of the head are good for judging size and distance
* sharp teeth are good for killing prey and tearing through flesh
* white fur provides good camouflage to hide from prey
* thick fur traps heat to keep warm
* lots of fat acts as a food store and helps to trap heat to keep warm
* strong legs are good for walking and swimming long distances
* sharp claws can grip ice
* hair on the soles of the feet helps provide a grip on icy surfaces.

4 The animals and plants that are more likely to survive are more likely to **reproduce**. If the features that helped them survive are genetically controlled they may be passed onto their offspring. Offspring who inherit those good features are in turn more likely to survive and reproduce than offspring that do not inherit the good features.

5 As this process continues over many generations, organisms with the **best** features will make up a **larger proportion** of the population and those with features that are **not as helpful** will gradually **disappear**. For example, if the fastest zebras in every generation survive, then over many generations zebras will get faster and faster.

■ Some populations of a species of animal can become separated, for example by sea or mountains. Natural selection means that each population group will evolve differently. Over many years each group may change so much that individuals from one group can no longer reproduce with individuals from the other. Each group has then evolved into a new species.

■ Natural selection explains why animals and plants are adapted to their surroundings. If the environment changes, the organisms that do not evolve and adapt may become **extinct**. There is more about adaptations in Unit 3.

QUESTION SPOTTER

▸ In questions about natural selection you will often be asked to describe the steps in the process.
▸ Although the examples you are given will vary, the steps themselves will be the same.

IDEAS AND EVIDENCE

▸ The theory of natural selection helps us to explain many scientific observations.

▸ An example is the issue of increased use of antibiotics and how this has led to the evolution of resistant bacteria.

The different coloured forms of the peppered moth mean there is a chance one form will be better camouflaged if the environment changes, so the species can survive.

■ The process of evolution has helped many organisms survive changes in their environment. A good example is the peppered moth and how it survived the effects of the Industrial Revolution in Britain. Before the Industrial Revolution the majority of peppered moths were light in colour. They were well camouflaged on the light-coloured tree trunks. The birds could not see them and they survived. Darker coloured peppered moths were not as well camouflaged and were eaten by the birds. During the Industrial Revolution many of the trees become black with soot. The darker coloured moths were then better camouflaged and survived better than the lighter coloured moths. In 1956 the Clean Air Act gradually reduced the amount of smoke pollution. After 1956 the numbers of lighter coloured moths gradually increased again as once more they had the better colour for camouflage.

CHECK YOURSELF QUESTIONS

Q1 a Why do we have a clearer idea of how vertebrate animals like mammals, birds and reptiles evolved than of how invertebrates like worms and slugs evolved?

b Why are there many gaps in the fossil record, even for animals with skeletons?

Q2 The long necks of giraffes are an adaptation allowing them to feed on the high branches of trees where many other animals cannot reach. Use the theory of natural selection to explain how giraffes could have evolved long necks from shorter-necked ancestors.

Answers are on page 152.

UNIT 10: EXAM PRACTICE

Exam tips

- **Read each question carefully**; this includes looking in detail at any **diagrams, graphs** or **tables**. Remember that any information you are given is there to help you to answer the question. Underline or circle the **key words** in the question and **make sure you answer the question that is being asked** rather than the one you wish had been asked!

- Make sure that you understand the meaning of the **'command words'** in the questions. For example:
  - **'Describe'** is used when you have to give the main feature(s) of, for example, a process or structure
  - **'Explain'** is used when you have to give reasons, e.g. for some experimental results
  - **'Suggest'** is used when there may be more than one possible answer, or when you will not have learnt the answer but have to use the knowledge you do have to come up with a sensible one
  - **'Calculate'** means that you have to work out an answer in figures.

- Look at the **number of marks** allocated to each question and also the **space provided** to guide you as to the length of your answer. You need to make sure you include at least as many points in your answer as there are marks, and preferably more. If you really do need more space to answer than provided, then use the nearest available space, e.g. at the bottom of the page, making sure you write down which question you are answering. **Beware of continually writing too much because it probably means you are not really answering the questions**.

- Don't spend so long on some questions that you don't have time to finish the paper. You should spend approximately **one minute per mark**. If you are really stuck on a question, leave it, finish the rest of the paper and come back to it at the end. Even if you eventually have to guess at an answer, you stand a better chance of gaining some marks than if you leave it blank.

- In short answer questions, or multiple-choice type questions, **don't write more than you are asked for**. In some exams, examiners apply the rule that they only mark the first part of the answer written if there is too much. This means that the later part of the answer will not be looked at. In other exams you would not gain any marks, even if the first part of your answer is correct, if you've written down something incorrect in the later part of your answer. This just shows that you haven't really understood the question or are guessing.

- **In calculations always show your working**. Even if your final answer is incorrect you may still gain some marks if part of your attempt is correct. If you just write down the final answer and it is incorrect, you will get no marks at all. Also in calculations you should write down your answers to as many **significant figures** as are used in the question. You may also lose marks if you don't use the correct **units**.

- In some questions, particularly short answer questions, answers of only one or two words may be sufficient, but in longer questions you should aim to use **good English** and **scientific** language to make your answer as clear as possible.

- If it helps you to answer clearly, don't be afraid to also use **diagrams** or **flow charts** in your answers.

- When you've finished your exam, **check through** to make sure you've answered all the questions. Cover over your answers and read through the questions again and check your answers are as good as you can make them.

EXAM QUESTIONS AND STUDENTS' ANSWERS (FOUNDATION TIER)

1 The diagram shows three plant cells.

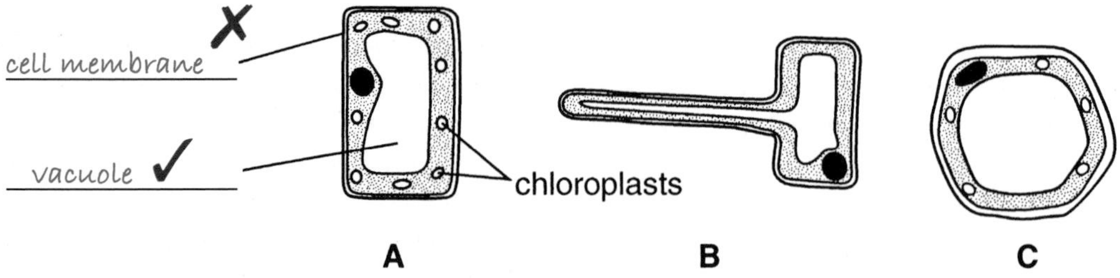

cell membrane ✗
vacuole ✓
chloroplasts
A
B
C

Each cell is from a different part of a plant.

a) Complete the labels on plant cell A. (2)

b) What is the job of the chloroplasts? (1)

chloroplasts absorb light ✗

c) Look at the cells. They are from a root, stem and leaf.

i) Which cell is most likely to come from a root? Choose A, B or C. Explain your answer. (2)

Cell B is from a root. Root cells do not have chloroplasts. ✓
Roots are below the soil and so can not absorb light. ✓

ii) Which cell is most likely to come from a leaf? Choose A, B or C. Explain your answer. (2)

Cell A is from a leaf. ✓ *This cell has chloroplasts, to absorb sunlight for photosynthesis.* ✓

d) Complete the word equation for photosynthesis. (2)

$$\text{Carbon dioxide} + \textit{water} \xrightarrow[\text{chlorophyll}]{\text{light}} \textit{glucose} + \text{oxygen}$$

✓ ✓

e) Plant cells need carbon dioxide for photosynthesis. Explain how carbon dioxide moves from the atmosphere into plant cells. (2 + 1)

Carbon dioxide enters the leaf through the stomata by diffusion. ✓ ✓ ✓

f) Plants make glucose by photosynthesis. Some of the glucose is turned into storage substances. Write down the name of one of these storage substances. (1)

starch ✓

11/13 (Total 13 marks)

How to score full marks

a) A common error is to become confused between cell wall and cell membrane. Try to remember, only plant cells have a cell wall, whereas all cells have a cell membrane. The cell wall is the outside part of the cell. The student gained one of the two marks available.

b) The student did not extend the answer to include chloroplasts absorb light for photosynthesis. Try to use scientific words. Include the word photosynthesis, even if you are not sure about the spelling. The examiner will know what you mean.

c) i) Cell **B** is most likely to come from a root. A root cell does not have chloroplasts; it is below the soil and so will not be able to absorb light. The student could have written about the shape of the cell. Root hair cells are elongated to provide a large surface area.

   ii) Cell **A** or **C** could have come from a leaf. Cells in the leaf need chloroplasts to absorb sunlight for photosynthesis. Make sure you study any diagrams and labels very carefully before you start to answer the questions.

d) If you are attempting the Foundation Tier paper, you will need to remember word equations for photosynthesis and respiration. If you are attempting the Higher Tier paper, you will also need to remember the symbol equations for photosynthesis and respiration.

e) The student could have extended the answer by describing how the carbon dioxide moves through air spaces in the spongy mesophyll layer. One mark each is awarded for the use of at least two of the following scientific terms: diffusion, concentration, stomata, mesophyll, permeable, membrane. When you see the symbol that indicates marks will be awarded for quality of the written answer (☞), you must check your answer. In this case, using correct scientific words would gain the extra mark. Read your answer and underline all scientific words in pencil. Check you have used them correctly.

f) Starch is the substance you will be most familiar with. However, other correct answers include fat, protein, oil, sucrose and cellulose.

---

2 The passage below is about Charles Darwin.

> ### Who Inspired Darwin?
> Thomas Malthus lived in the early 19th century. He wrote 'An Essay on the Principle of Population'. In this essay he pointed out that human beings produced far more offspring than ever survive. However, the adult population tends to remain stable from generation to generation.
>
> Darwin realised that this idea applies to other animals. For example one fish, which lays thousands of eggs in a year, would over-populate an area with its offspring if they all survived.
>
> The work of Malthus helped Darwin to develop his own ideas of how a species changes. He produced his theory of natural selection. Darwin realised that there must be a reason why some offspring survived but others did not. He suggested that small variations between individuals of a species might give certain individuals a better chance of survival. For example, those organisms with characteristics that made them better at escaping from predators or finding food would have a better chance of surviving.

a) i) What is meant by the phrase 'the adult population tends to remain stable from generation to generation'? (2)

The numbers remain constant over a long period of time ✓ ✓

ii) Suggest why fish lay thousands of eggs rather than just a few. (2)

Fish have many predators so if only a few eggs were laid not many would survive. There would not be enough to continue the species ✓ ✓

iii) What can cause 'small variations between individuals of a species'? (1)

Small variations can be caused by mutations ✓

iv) What is meant by the phrase **natural selection**? (3 + 1)

Some individuals have characteristics that make them more successful. These individuals survive better in their environment. The successful characteristics are passed on to the offspring. ✓ ✓

b) Here are four statements about evolution. Tick the box beside the statement that is false. (1)

The theory of evolution was developed by Darwin. ☐

DNA is the genetic material that transfers information from generation to generation. ☐

Acquired characteristics **cannot** be passed on from parent to offspring. ☐

Nature plays an important part in artificial selection. ☑ ✓

c) Suggest **two** ways that scientists can let other groups of scientists know about their ideas. (2)

1 television ✗

2 publishing journals or books ✓

(Total 12 marks)

10/12

☐ How to score full marks

This is an Ideas and Evidence-style of question. Do not worry that you may not have been taught about everything in the passage. The questions ask about your understanding of Biology.

a) i) Full marks have been awarded. The student correctly stated the numbers 'remain constant' and then extended the answer to state 'over a long period of time'. It is easy to miss the second mark. Always look at the number of marks available and check your answer has the same number of important points.

   ii) The two marks have been awarded. The student has given a clear explanation.

   iii) The student gained a mark for stating a mutation. Environmental factors could have been an alternative answer.

   iv) The student gained three out of the four marks available. When a question is worth four marks, it is important to plan your answers. It is too easy to forget to write about one important fact. It is a good idea to list four key words in the margin. As you write about each key word you can cross it out.

The student gained a mark for explaining that individuals with beneficial characteristics survive and pass on the characteristics to offspring. The fourth mark could have been gained by explaining about how the environment or competition creates pressure on individuals. The student also failed to write that those individuals that survive are able to reproduce.

A mark was available for the quality of the written answer, which here was for accurate spelling, punctuation and grammar.

b) The student correctly ticked the box. **If you are not sure, make a guess.** A well-informed guess might gain a mark, leaving an answer blank will not gain any marks.

c) The answer of a television is too vague to gain a mark. In the second part of the question the student gains one mark for a more detailed attempt.

I Use a word or phrase from the box to complete each sentence.

The first one has been done for you.

increases	decreases	stays the same

After injecting with a used needle, the chance of getting hepatitis A   *increases*   .

After taking an antibiotic, the number of disease-causing micro-organisms in the body (1)

After taking heroin, the amount of pain felt (1)

For regular smokers, the chance of developing lung cancer (1)

When a person is healthy, the number of white blood cells (1)

2 A runner might drink a special 'sports drink' at intervals during a marathon race. The table shows the substances present in a sports drink.

Substance	Percentage
Water	
Sugar	5.0
Ions	0.2

a) Complete the table to show the percentage of water in the sports drink. (1)

b) The runner sweats and also breathes heavily during the race.

i) Why does the runner need to sweat? (1)

ii) Which **two** substances in the table are lost from the body in sweat?

(1)

iii) Which substance in the table is lost from the body during breathing? (1)

c) How does the sugar in the sports drink help the athlete during the marathon? (2)

Answers are on page 153.

I   The diagram shows the human heart.

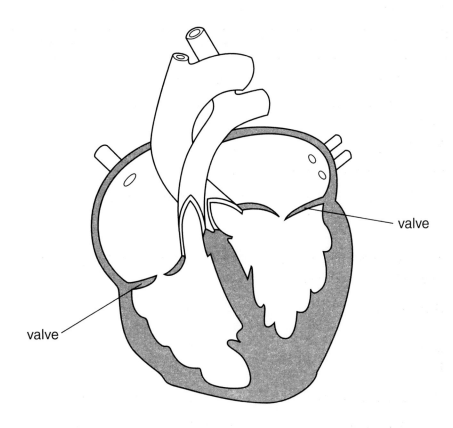

valve

valve

a)   Write down the job of the heart.   (1)

to pump blood around the body ✔

b)   There are two sets of valves shown in the diagram.

    i)   Write down the job of these valves   (1)

    to stop the blood from flowing backwards ✔

    ii) Describe **how** these valves carry out their job.   (2)

    they open and close ✗

    iii) Where else in the circulatory system are valves found?   (1)

    in between the atrium and the ventricle ✗

c) This is a picture of William Harvey.

He was an English doctor who lived from 1578 to 1657.

He made important discoveries about the circulation of blood.

Before Harvey's ideas on blood were known, people believed the ideas of a scientist called Galen.

i) Describe how William Harvey's ideas on circulation differed from Galen's. (2)

Harvey demonstrated to people that blood flowed around
the body ✔

ii) Suggest two reasons why Harvey's ideas spread faster than Galen's. (2)

1 _More books and papers were published during Harvey's time_ ✔

2 _Not many things were published during Galen's time_ ✗

(4/9) (Total 9 marks)

How to score full marks

Always take time to read the words above diagrams. Your eyes will automatically go straight to the diagram and then move down the page. You may miss something very important. Try and train yourself to start at the top of each page. Then study the diagram, reading all labels. Make sure you understand the diagram before you start reading the questions.

a) The student has gained one mark for a complete statement. A vague answer such as 'it pumps blood' would not have been sufficient to gain a mark. Always try to write as much detail as possible.

b) i) The student has demonstrated a clear understanding of the job of the valves.

ii) The student has some understanding of how valves work but this vague answer does not get a mark. An acceptable answer would be 'the valves are forced shut by the pressure of the blood'. This part of the question has two marks. For the second mark the student needs to say 'tendons stop the valves from inverting'.

iii) The student has not read the question carefully. 'Where else in the circulatory system' does not include the heart. The correct answer would be 'inside veins'.

c) This is an example of an Ideas and Evidence-style question.

i) Always look at the number of marks available for an answer. The number of lines provided to write your answer is also an indication of how much you should write. The student should have given two pieces of information to get full marks.

The student could have extended the answer by writing about any of the following. Harvey realised that different types of vessels carried blood around the body. Harvey demonstrated the existence of valves in veins. Harvey suggested there was a double circulation; he did not believe that blood just passed directly from one side of the heart to another. Harvey also predicted the existence of capillaries. This could not be supported till many years later when microscopes were developed.

ii) The student gained a mark for the first part of the answer. However, a second mark cannot be given for the same point written in a different way. The idea about communication of scientific knowledge is a very important issue for Ideas and Evidence. The student could have stated that more people could read during Harvey's time. A mark could also have been given if the student had discussed the availability of more widespread travel.

2 The bucardo is a type of goat that lived high in the mountains of Spain.

The bucardo had thick fur and sharp, curved hooves.

a) Explain how these features helped the bucardo to live high in the mountains. (2)

thick fur _because it is cold on the top of the mountain_ ✗

sharp curved hooves _to fight off other animals_ ✗

b) The following passage about the bucardo appeared in a recent newspaper. Read the passage carefully and use it to help you answer the questions.

> **Yet another extinct animal**
>
> The last known survivor of a species of a mountain goat called the bucardo was found dead on January 6th 2000.
>
> The female bucardo had been crushed under a tree. The bucardo had been declared a protected species in 1973 because of its low numbers.
>
> This species of mountain goat was once widespread but numbers had decreased due to various actions by man and natural disasters such as landslides.

i) Various actions by man had caused a decrease in the number of bucardos. Suggest what one of these actions may have been. (1)

landslides ✗

ii) Explain how making a type of animal a protected species helps it to survive. (2+1)

👉 because it is a protected species people can not destroy its habitat ✔

iii) The bucardo lived high in the mountains. Suggest why this made it a difficult species to protect. (1)

It is a difficult species to protect because it is high up in the mountains ✗

c) In 1999, a tissue sample was taken from the ear of the last bucardo and frozen. Scientists want to try to clone the bucardo using these cells. They will try to create a bucardo embryo by using the nuclei from the stored cells. (2)

i) What is a clone?

an identical copy ✔

ii) If the scientists do manage to produce a bucardo embryo, what problem will they face next? (1)

The clone will only produce females ✔

d) It is possible to produce clones of human body cells. Explain how this may be useful and why some people object to this process. (3+1)

👉 One useful process would be to produce clones of humans to create embryos for infertile couples ✔ Some people object because it is against their religious belief. ✔ ✔

(6/14) (Total 14 marks)

☐ How to score full marks

The story of the bucardo provides a storyline for the question and helps to make the question more relevant to the real world. Some of the questions on your exam paper will have an unfamiliar storyline. This does not matter; the questions are asking about your knowledge of the biology you have been taught.

a) This does not answer the question. The student's comment is not an explanation of how the thick fur helped the bucardo to live high in the mountains. The examiner cannot assume the student knows the answer so no mark can be given. A mark could only be given for a comment about how the thick fur provides insulation to keep the bucardo warm.

The student has not read the question. It seems the student has seen the horns in the picture and assumed that the question is about them. The correct answer would have been to write that the hooves would help the bucardo's grip on the rocks.

b) There is a lot of information in this part of the question. Always read very carefully all the information you are given in a question.

i) The student has not read the question carefully. The correct answer would have referred to an activity of humans, such as poaching, hunting, destruction of habitat or removal of the bucardo's food supply.

ii) Always look at the number of marks available for an answer. The student should have given two points in the explanation. A mark could have been awarded for comments about laws made to make hunting illegal or controlled, the introduction of breeding programmes or public awareness programmes.

There was an extra mark available for the quality of the written answer, which here would have been for using correct spelling,

punctuation and grammar. Unfortunately this student failed to gain the mark because it was not a well-structured sentence. **Where you see a pencil symbol take time to check your answer.**

iii) Rewriting the question is a common mistake in exam. Clearly this student has not taken the time to understand the question before starting to write the answer. If you are not sure what the question is asking, read it again. If necessary leave it and return to it later. The correct answer would have referred to the difficulty of watching for poachers or monitoring the bucardo because of difficult access. It is also difficult to protect the bucardo from natural disasters such as landslides.

c) i) The student has gained one mark for the idea of an identical copy. For the other mark the student should have included the term 'genetically', i.e. a 'genetically identical copy'. Always look at the number of marks available for an answer.

ii) The student has demonstrated a clear understanding of the problems. Other comments worth marks could have been about the need to find an animal to implant the embryo into or there being no males to mate with. Comments about lack of genetic variation would also gain a mark.

d) Always look at the number of marks available for an answer. The student gave one use and one objection. The third mark could have been awarded for a comment about cloning of organs for organ transplants, which would not be rejected by the host. Objections could have included ethical reasons.

The extra mark for the quality of the written communication was awarded because the student wrote clearly about a use and an objection.

1 Read the following passage that appeared in a recent newspaper.

> **Insulin pill could free diabetics from injections**
>
> About a million people in Britain are thought to need insulin injections. These people have a condition called diabetes. They have to inject themselves up to three times a day.
>
> Scientists have been trying for years to develop an insulin pill that diabetics can swallow, rather than having injections. The main problem has been the fact that insulin is a protein and so is easily destroyed in the stomach. Now experts have been able to coat the tablets with a special molecule, which stops the insulin being destroyed.
>
> A doctor said 'The beauty of the tablet is that when the insulin is absorbed into the blood stream it passes straight to the liver.'

a) Why do diabetics need insulin injections?   (2)

b) Diabetics inject themselves up to three times a day rather than giving themselves one large injection. Suggest **one** reason for this.   (1)

c) Why is insulin destroyed in the stomach if swallowed?   (2)

d) The doctor was pleased that the insulin form the pill was passing straight to the liver. Why does the insulin need to reach the liver?   (2)

2 In the eighteenth century, surgeons did not wear special clothing or wash their hands before operations. Many of their patients died from infections.

a) Suggest why patients often died from infections after operations.   (1)

b) In the nineteenth century, Joseph Lister told surgeons to use sprays of carbolic acid in operating theatres and to wash their hands. The graph shows the effect that using Lister's instructions had on the number of patients who died from infections after surgery.

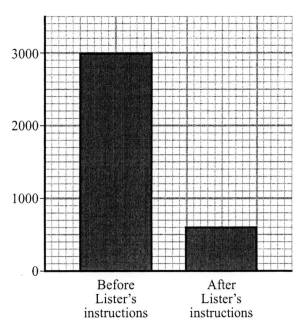

Number of patients dying from infections per 10 000 operations

Describe how Lister's instructions affected the number of patients dying from infections after surgery. (2)

c) Kidney transplants were introduced in the twentieth century as one way of treating patients with kidney failure.

    i) Give **one** other way of treating kidney failure. (1)

    ii) The patient's body may reject a transplanted kidney unless doctors take precautions. Some of these precautions are listed below.
A donor kidney is specially chosen.
The recipient's bone marrow is treated with radiation.
The recipient is treated with drugs.
The recipient is kept in sterile conditions.

    Explain how each of these precautions may help the patient to survive. (4)

Answers are on pages 153–154.

1 The flow chart shows some of the stages in making beer.

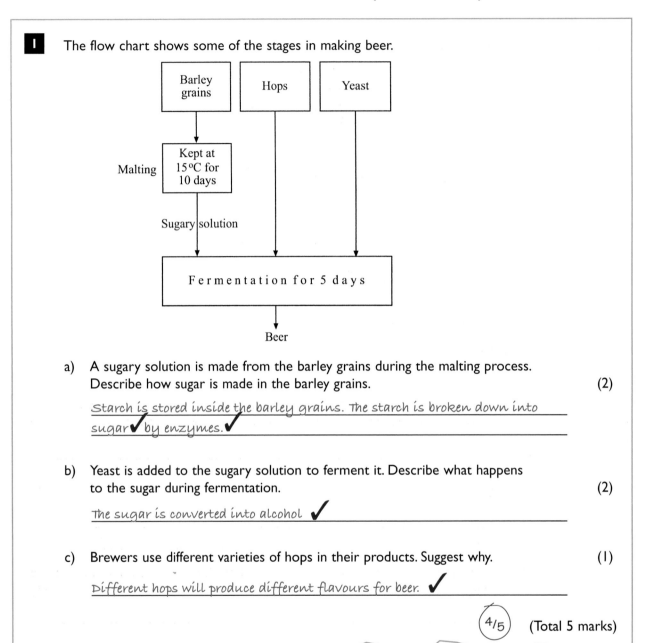

a) A sugary solution is made from the barley grains during the malting process. Describe how sugar is made in the barley grains. (2)

> Starch is stored inside the barley grains. The starch is broken down into sugar ✔ by enzymes. ✔

b) Yeast is added to the sugary solution to ferment it. Describe what happens to the sugar during fermentation. (2)

> The sugar is converted into alcohol ✔

c) Brewers use different varieties of hops in their products. Suggest why. (1)

> Different hops will produce different flavours for beer. ✔

4/5 (Total 5 marks)

⬚ How to score full marks

a) Both marks have been awarded. The student has clearly described how barley grains made sugar.

b) Always look at the number of marks available for an answer. The student gained one of the two marks available. The student could have gained the other mark for stating that fermentation is a type of anaerobic respiration. A mark could also have been awarded for stating that part of the sugar produces carbon dioxide.

c) This question says 'suggest', so the student has to apply knowledge to the topic. The student gains a mark for the answer of different flavours. A mark could also have been given for a comment about different hops causing different bitterness of taste.

2 The drawing shows a cactus plant.

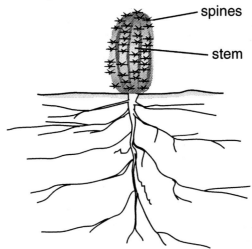

spines

stem

This cactus plant lives in a hot, dry desert.

The leaves have evolved to become spines.

The stem is green.

a) i) Plants have roots to take in water. This cactus plant has very long roots. Suggest how very long roots help cactus plants to survive in dry desert conditions. (1)

Long roots can absorb more water ✔

ii) Describe two other jobs of roots. (2)

1 *to hold the plant inthe ground* ✔

2 *to take in minerals from the soil* ✔

iii) Write down the name of the transport tissue that carries water from the roots to the rest of the plant. (1)

xylem ✔

b) Is a cactus plant a producer or a consumer? Explain your answer. (1)

The cactus plant is a producer because it can photosynthesise to produce its own food ✔

c) The spines (leaves) of a cactus plant are not very efficient at photosynthesis. Explain why. (2)

They have a small surface area ✔

✗

(6/7) (Total 7 marks)

☐ How to score full marks

a) i) A mark would be given for any one from: long roots can absorb more water or absorb water faster; large roots can collect water from a large area; they can compete for water and can help the plant to survive when water is scarce; large roots can store water. A common error is to say that roots take in food. Do not forget the important fact that green plants have to make their own food.

ii) A mark each would be given for any two from: anchorage; water storage; take in minerals. The question asks for two **other** jobs: check that you do not write a fact that has already been given.

iii) It is easy to confuse xylem and phloem. Try to remember that x and w are close together in the alphabet, and xylem carries water; phloem carries the food (glucose) for the plant.

b) This answer is worth one mark. The mark is for the explanation, not for the answer 'producer.' Marks are not awarded for a choice from two, such as producer or consumer.

c) A mark each would be given for any two from: they have a small surface area; less light can be absorbed; they have less chlorophyll; they have very few stomata; very little carbon dioxide is taken in. The student only gained one of the two marks available. The student only wrote one comment. The question did not ask for two comments, but two marks were allocated to it. Always check the marks awarded, in brackets at the end of the lines. Two marks will require two pieces of information.

QUESTIONS TO TRY (HIGHER TIER)

1 This question is about response to stimuli.

a) The diagram shows the structure of a motor neurone.

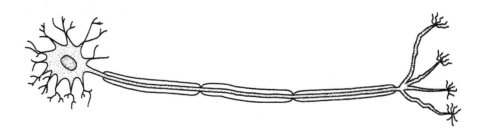

i) Draw an arrow on the diagram to show the direction that a nerve impulse usually travels. (1)

ii) Motor neurones carry nerve impulses from the central nervous system to the muscles. Write down **two** ways in which the neurone is adapted to do this. (2)

b) Responses in the body can also be brought about by hormones.

i) How do hormones reach their target in the body? (1)

ii) The type of response brought about be hormones is often different from the responses brought about by neurones. Write down **two** differences. (2)

c) Plants also respond to stimuli. Liz does an experiment to measure the growth of a plant shoot. She shines a light at the shoot from the side. She measures the length of the side of the shoot nearest the light. She also measures the length of the shaded side. Liz does this for several days.

    i) Compare the growth on each side of the shoot after the light is turned on. (3+1)

    ii) Explain how these changes in growth are thought to happen. (3)

    iii) Explain the advantage to the plant of this type of response. (2)

2 Doctors are concerned that people who eat snacks all the time may be damaging their health. The high levels of insulin in their blood could lead to diabetes.

a) Eating a lot of small meals through the day leads to a steady flow of glucose **into** the blood. Explain why. (1)

b) The level of glucose in the blood is controlled by insulin. Insulin is a hormone produced by the pancreas. What is a hormone? (1)

c) Glucose in muscle tissue can be changed into an insoluble polymer. Insulin causes the change of the glucose into a polymer.

    i) What is the name of the polymer? (1)

    ii) Describe how insulin travels from the pancreas to the muscle tissue. (1)

d) Muscle tissue transfers energy in glucose into useful physical work. The brain instructs muscle tissue to do this.

    i) Describe how this instruction travels from the brain to the muscle tissue. (1)

    ii) Complete and balance this symbol equation for the aerobic respiration of glucose in muscle cells. (1)

$$C_6H_{12}O_6 + 6O_2 \rightarrow 6CO_2 + $$

    iii) Muscle tissue can also use anaerobic respiration to do useful physical work. State **two** reasons why aerobic respiration is normally used. (2)

Answers are on pages 154–155.

ANSWERS

Unit 1: Life processes and cells
1 Characteristics of living things (page 1)

Q1 a It does not contain any chloroplasts (and therefore no chlorophyll).

Comment Any green parts of a plant will contain chloroplasts.

b Onions grow underground, so they will not receive any light. There is no point in them having chloroplasts as they will not be able to photosynthesise.

Comment By going into detail and mentioning photosynthesis you show clearly that you understand the reason.

c It has a cell wall, a large vacuole and a regular shape.

Comment There are some exceptions, in that some plant cells may not have large vacuoles or a regular shape. But they all have a cell wall.

Q2 a Mitochondria.

Comment This is a better answer than simply saying cytoplasm.

b Cell membrane.

Comment Don't forget that everything that goes in or out of a cell goes through the membrane.

c Nucleus or chromosomes.

Comment Both are correct. Chromosomes is a more specific answer and shows more understanding.

d Vacuole.

Comment Don't forget that only plants have large vacuoles.

e Cell wall.

Comment The membrane would simply burst if the cell became too big. The cell wall is more rigid and resists this. You will find out more about this in Unit 6.

Q3 a Organism.

b Cell.

c System.

d Organ.

e Tissue.

f Organ.

Comment Organisms are whole animals or plants. (Don't forget that there are some organisms, like bacteria, which are only one cell big.) Organs are the separate parts of a body, each of which has its own job(s). Organs are usually made of different groups of cells (tissues).

2 Transport into and out of cells (page 4)

Q1 The salt forms a very concentrated solution on the slug's surface so water leaves its body by osmosis.

Comment This happens with slugs because their skin is partially permeable. It would not happen with us because our skin is not.

Q2 a Diffusion.

Comment There is a lower concentration of carbon dioxide inside the leaf as it is continually being used up.

b Neither.

Comment The food is squeezed along by muscles in the gullet.

c Neither.

Comment Again, water is being forced out.

d Osmosis.

Comment As the celery is dried up the cells contain very concentrated solutions.

Q3 To provide energy for the active transport of minerals like nitrates into the roots.

Comment The concentration of minerals in soils is usually lower than the concentration of minerals in plant cells. Plants therefore can't rely on diffusion to get more of these minerals and must expend energy on actively 'pulling' them in.

Q4 Diffusion is the movement of particles from an area of higher concentration to an area of lower concentration. In active transport particles move against a concentration gradient. They move from an area of low concentration into an area of high concentration. Active transport requires energy from respiration. Examples of diffusion include: movement of oxygen from lungs into blood; movement of carbon dioxide from blood into oxygen; movement of glucose from villi in small intestines into blood. Examples of active transport include: the movement of minerals from the soil into the root hair.

Comment Questions about active transport are restricted to Higher Tier exam papers. The important facts to include in your answer are:
• particles move against a concentration gradient;
• energy (from respiration) is needed.

UNIT 2: HUMAN BODY SYSTEMS

1 Cells and respiration (page 8)

Q1 To provide energy for life processes like movement and to generate warmth.

Comment Don't give vague answers like 'to stay alive'. If a question in an exam has several marks then give the correct number of points in your answer.

Q2 In every cell of the body.

Comment You could be even more specific and point out that respiration occurs in the cytoplasm or in the mitochondria. Many students confuse respiration and breathing, and would have said 'lungs'.

Q3 a Aerobic respiration provides more energy (for the same amount of glucose) and does not produce lactic acid (which will have to be broken down).

Comment Given sufficient oxygen, respiration will be aerobic.

b If more energy is needed and the oxygen necessary cannot be provided.

Comment Anaerobic respiration is not as efficient as aerobic respiration. It is only used as a 'top up' to aerobic respiration.

2 Blood and the circulatory system (page 10)

Q1 Substances, for example oxygen, will quickly diffuse into the centre of a small organism. For larger organisms, diffusion would take too long so a transport system is needed to ensure quick movement of substances from one part of the organism to another.

Comment Another way of explaining this is to say that smaller organisms have a larger surface area to volume ratio (or a larger surface relative to their size).

Q2 a In the lungs.

Comment The blood arriving at the lungs contains little oxygen and is there to collect more.

b In respiring tissues around the body.

Comment Every cell will need oxygen. Active cells like those in muscles will need most.

c To ensure that their bodies collect enough oxygen.

Comment *If the air contains less oxygen then less oxygen will enter the blood unless the body has some way of compensating.*

Q3 a The ventricles pump the blood but the atria simply receive the blood before it enters the ventricles.

Comment *The ventricle walls contract and relax to squeeze out blood and take more in. This takes a lot of muscle. The atria, by comparison, do not need to squeeze as hard and must be easily inflated by the incoming low-pressure blood.*

b The left ventricle has to pump blood around the whole body (apart from the lungs). The right ventricle 'only' sends blood to the lungs.

Comment *This is why your heart sounds louder on your left side.*

Q4 a Veins have valves, thinner walls and a larger lumen.

Comment *The veins contain blood at a reduced pressure and are built so that they don't resist the flow of blood but rather help it on its way to the lungs.*

b Capillary walls are permeable, allowing diffusion. Arteries and veins do not have permeable walls.

Comment *The arteries carry blood quickly to each organ or part of the body and the veins bring it back. The capillaries form a branching network inside organs.*

3 Air and breathing (page 15)

Q1 a Just two, the wall of the alveolus and the wall of the blood capillary.

Comment *This short distance is important so gases can easily move between the alveoli and the blood. Another way of looking at it is that the short distance increases the concentration gradient.*

b To ensure that oxygen rapidly diffuses from the area of high concentration, in the alveoli, to the area of low concentration, in the blood.

Comment *Don't forget that there is also a concentration gradient for carbon dioxide.*

Q2 The diaphragm lowers and the ribcage moves upwards and outwards.

Comment *Both of these changes increase the volume of the thorax.*

Q3 To keep them open to ensure that air can move freely.

Comment *This is particularly important because when the air pressure drops to take in more air from the outside this could otherwise make the air passages close in on themselves.*

4 Food and digestion (page 19)

Q1 a Proteins are too big to pass through the wall of the ileum, so they have to be broken down into the smaller amino acids.

Comment *Only molecules that are already small do not need to be digested.*

b Physical digestion is simply breaking food into smaller pieces. This is a physical change. Chemical digestion breaks down larger molecules into smaller ones. This is a chemical change because new substances are formed.

Comment *Many students, if asked to describe digestion, lose marks because they do not describe the breakdown of **molecules**.*

c They are denatured.

Comment *This means they change their shape irreversibly, and can not work. In an exam **do not** say that they are 'killed' because they were never alive.*

d To neutralise the acid from the stomach, because the enzymes in the small intestine can not work at an acidic pH.

Comment Many students think that stomach acid itself breaks down food. It is the fact that it allows enzymes in the stomach to work that is important.

Q2 a Ingestion is taking in food. Egestion is passing out the undigested remains. Excretion is getting rid of things we have made.

Comment Many students lose marks because they confuse egestion and excretion. Passing out faeces is egestion. Getting rid of urea in urine or breathing out carbon dioxide are examples of excretion because these substances are made in the body.

b Ingestion: mouth. Egestion: anus.

Comment Excretion occurs in various places. You will find out more in Unit 4.

Q3 They provide a large surface area. They have thin permeable walls. They have a good blood supply causing a concentration gradient. They are both moist.

Comment The similarities are not a coincidence. These factors are vital to ensure the efficient absorption of the different substances.

UNIT 3: ANIMAL ADAPTATIONS
1 Feeding adaptations (page 25)

Q1 The housefly has a proboscis that uncurls and moves into the food. The food can then be sucked up the proboscis.

Comment Insects are so small they do not have room for a complex digestive system. Insects have to digest their food outside of their body then suck the partially digested food into their body.

Q2 The larvae and adult insects feed on different foods from different habitats. This ensures that each stage is not competing the other for food.

Comment Do not forget that each stage of the insect life cycle is very different.

Q3 a (i) Incisors have sharp edges to bite or tear food.

Comment Read the whole of the question first before you start to write. This will ensure you do not write too much, including answers to the next part of the question.

(ii) Molars have a flat upper surface to crush and chew food.

Comment Imagine biting and chewing on an apple. This will help you to think about the different functions of your teeth.

b Three from: lions have sharper incisors; lions have larger sharp, pointed canines; lions' premolars and molars include special large carnassial teeth; sheep have no canine teeth; sheep have flat upper surfaces to their molars and no carnassial teeth.

Comment You need to think about the different diets of each animal before you start to write about their teeth.

2 Adaption for habitats (page 28)

Q1 a (i) Tendons

Comment A common error is to confuse ligaments and tendons. It will help to remember the letter 't' appears in the words tendon and attach. This will help you to remember that tendons attach muscle to bone.

(ii) Cartilage

Comment Cartilage also reduces friction due to its smooth slippery surface.

(iii) Ligaments

Comment Ligaments are made up of strong fibres that help them to carry out their function.

b The fingers perform more movements than the arm. The fingers make finer movements than the arm.

Comment Remember that bones do not move. Movement occurs at a joint. The more movement required will mean more joints are needed.

Q2 **Three** from: a cross-section of a bird's wing has a curved shape; called an aerofoil; this shape creates a lower pressure above the wing; and so generates lift.

Comment It is difficult to know how many points to include in your answer. Check the number of marks that are to be awarded for the answer. These are shown in brackets at the end of the question.

Q3 **a** Produce wave-like movements of the body.

Comment Muscles can only pull, they cannot push. This means they have to work in pairs.

b Provides a pushing force backwards against the water to move the fish forwards.

Comment This is the reason why divers need to wear flippers on their feet, when swimming under water.

c Reduces resistance.

Comment Boats are built to a streamlined shape for the same reason.

UNIT 4: HUMAN BODY MAINTENANCE
1 The nervous system (page 31)

Q1 **a** A change, either in the surroundings or inside the body, that is detected by the body.

Comment For example, if you were crossing a road and noticed a bus approaching the stimuli would have been the sight and sound of the bus.

b Receptors detect (sense) stimuli. The sense organs are receptors. Effectors are the parts of the body that respond. Effectors are usually muscles but can also be glands.

Comment Some students get confused between effectors and the 'effects' that they produce.

Q2 **a** Brain and spinal cord.

Comment Many reflexes do not need the brain in order to function and the spinal cord is therefore not simply a nerve pathway to and from the brain but also processes some information itself.

b Neurones are nerve cells. A nerve in the body is usually a bundle of neurones wrapped in connective tissue.

Comment In exam questions you may pick up extra marks for using the correct terms.

c Membrane, cytoplasm, nucleus.

Comment Although neurones are specialised they are still cells and so share these features with other cells.

d **Two** from: cell body, nerve fibres (axons or dendrons), nerve endings or dendrites.

Comment If you are asked for a certain number of answers in an exam only answer up to that number as any more will not be marked.

Q3 **a** A and C.

Comment The others do involve some kind of response, but reflexes happen **automatically** – without thought.

b Since they do not have to be thought about, reflexes happen very quickly. This means that if there is some kind of danger you can respond very quickly to reduce or prevent any harm.

Comment It is possible to override some reflexes by thinking about it, so you could hold on to something hot even though that is not usually a very wise thing to do.

2 The endocrine system (page 37)

Q1 In the nervous system, signals are sent very quickly as electro-chemical impulses along neurones and the effects of the signal usually last a short time. In the endocrine system signals are sent more slowly as hormones through the blood and the effects usually last longer.

Comment Both of these systems are involved in co-ordination (responding to stimuli). Both are needed because of the different nature of their effects.

Q2a Fighting and running away both need energy. This is released by respiration, which uses oxygen to release energy from glucose. So breathing increases to take in more oxygen, and glucose is released from stores in the liver and muscles. The blood is redirected and heart rate increased to carry the oxygen and glucose quickly to the muscles that need it. Sweating starts because the body will need cooling. Hairs stand up and pupils dilate to make an animal look larger and more fierce.

Comment Although an exam question may be apparently about one particular topic, to gain full marks you might have to include ideas from other topics. In this example ideas about respiration are relevant. This is likely to happen in questions that require extended answers.

b Otherwise all the effects caused by adrenaline would persist.

Comment Where hormones are involved, constant fine adjustments need to be made – for example, in the amounts of insulin required in the blood. If hormones are not being constantly broken down and at the same time secreted this balance could not be maintained.

Q3 Glucose and glycogen are both carbohydrates, but glucose is made of small molecules that dissolve in the blood whereas glycogen is made of larger molecules (lots of glucose molecules joined together) that are insoluble and do not travel through the blood. Glucagon is a hormone that converts glycogen back into glucose (i.e. it does the opposite of insulin).

Comment It is very common for students to confuse these – especially glycogen and glucagon.

3 Homeostasis (page 41)

Q1 The farmer will have been sweating to stay cool and so will have lost water this way. Unless the farmer has drunk a lot, his or her kidneys will have to reduce the amount of water that is lost as urine. This is why there is less. However, there will be just as much urea, so the urine is more concentrated and darker in colour.

Comment The body has to lose extra water by sweating to maintain body temperature. Humans cannot reduce how much is lost in breathing as the lining of the alveoli in the lungs is always moist, nor can the amount lost in faeces be reduced greatly. The only way therefore to cut down on losses is to reduce urine output.

Q2 They remove carbon dioxide and water from the body.

Comment Excretion is the removal of substances that have been produced in the body. Carbon dioxide is a waste product of respiration, as is some of the water.

Q3 The veins would have a lower concentration of salt and urea than the arteries. Also, as with any other organ, the oxygen level and blood pressure would be lower.

Comment Although most other substances in the blood (e.g. glucose) are also initially filtered from the blood, most are the absorbed back before the blood leaves the kidneys. Also, don't forget that the arteries are carrying blood away from the heart into the kidneys, and the veins the reverse.

4 The immune system (page 46)

Q1 a Jenner noticed that milkmaids suffering with cowpox did not catch smallpox. He took pus from a cowpox sore and infected a boy with cowpox. He left the boy for several weeks. Jenner then took pus from a smallpox sore and infected the boy with small pox.

Comment *Be prepared to answer an Ideas and Evidence-style question that might ask about the risks Jenner took when carrying out this procedure. If the young boy had died Jenner risked being accused of murder.*

b Four from: the vaccine contains a mild or dead form of the pathogen; when the vaccine is in the blood the white blood cells produce antibodies to attack the pathogen; antibodies remain in the bloodstream for a long time; when live pathogens later enter the body the antibodies are still in the bloodstream; the antibodies destroy the pathogens.

Comment *It is important to state the vaccine contains a dead or mild form of the pathogen.*

c Four from: passive immunity is when you are injected with ready-made antibodies; you may already have the pathogens inside your body; so you cannot wait for the body to produce its own antibodies; the injected antibodies prevent you from suffering from the disease.

Comment *The difference between active and passive immunity is often confused. Make sure you can write correctly about each during your revision. Your thinking needs to be very clear whilst in the exam room.*

Q2 a HIV is the name of the virus that invades the white blood cells. AIDS is the condition that develops due to HIV.

Comment *This is a common misunderstanding. Take care when reading about this topic: in some newspapers and magazines the two abbreviations are often confused.*

b Three from: people using a condom during sexual relations; not having may sexual partners; drug-users using clean hypodermic needles.

Comment *When asked for three ways make sure you only list three. The examiner will mark the first three and ignore the rest.*

c Patients with AIDS have very few white blood cells, so the immune system becomes very weak. Patients develop pneumonia because they no longer have any immunity for it.

Comment *The question could have stated any illness. The important fact is the immune system is weakened.*

Q3 a Red blood cells are needed in large quantities to carry oxygen, whereas white blood cells are fighting infection and most of the time we are not recovering from diseases.

Comment *Remember to relate the blood cells to their function to enable you to answer the question fully.*

b When there is an infection to fight off.

Comment *Even when we are making lots of white blood cells to fight an infection there are many times more red blood cells.*

Unit 5: Health and disease
1 Types of disease (page 49)

Q1 Four from: the virus has a protein coat that contains DNA; virus injects its DNA into a cell; the DNA takes control of the cell and produces a new set of instructions for the cell; the cell makes hundreds of copies of the virus; the cell dies and releases the viruses.

Comment *This type of question will be worth four marks. Check the number of marks available and then plan your answer. Write four key words in the margin to help you structure your answer.*

Q2 a Infectious diseases are caused by microbes and can be caught from someone else. Non-infectious diseases are caused by our genes or by the environment and are not caught from someone else.

Comment Try to include scientific terms. Vague comments such as 'non-infectious diseases cannot be passed on' will not gain marks.

b (i) One from: whooping cough; tetanus; food poisoning; diphtheria; typhoid; athlete's foot; ringworm; common cold; influenza; chicken pox; AIDS; smallpox; amoebic dysentery; malaria; sleeping sickness.

Comment Try to remember a few examples of each. You may have had some of these infectious diseases when you were younger.

(ii) One from: cancer; silicosis; asbestosis; coronary heart disease; dietary disease such as kwashiorkor disease; hormonal malfunctioning such as thyroxin; inherited diseases such as cystic fibrosis; Huntington's chorea, Down's syndrome, sickle-cell anaemia and haemophilia.

Comment Another question might ask for inherited diseases. Make sure you can remember a few of each type of non-infectious disease.

Q3 a Virus, bacteria, protozoa, fungi, parasite.

Comment There are always exceptions to every rule. Some parasites could be smaller than some fungi. Try to consider the majority when answering this style of question.

b Answers can be taken from the table and information in Revision Session 1 (pages 49–50).

Comment It might be easier to write your answer as a list or table. This is acceptable and often easier for your examiner to mark. First check the question does not carry a quality of written communication (QWC) mark. Any question that does carry a QWC mark must have an answer written in sentences and include scientific terms.

2 The spread of disease (page 51)

Q1 Three from: if the public understands the problems and reasons; disease and ill health will be reduced; less treatments will be required; less money will need to be spent on treatments.

Comment This answer is not easy to put into words. Practise writing this type of extended answer. Check your answers by underlining the parts that can be awarded a mark.

Q2 a Mosquito (*Anopheles*).

Comment If you are attempting the Foundation Tier exam you will be expected to remember the mosquito transmits malaria. If you are attempting the Higher Tier exam you may be expected to remember that it is the female *Anopheles* mosquito.

b Inside red blood cells.

Comment A vague comment such as 'in the blood' will not gain a mark.

c Four from: the mosquito introduces *Plasmodium* into red blood cells; the *Plasmodium* grows inside the blood cells; the *Plasmodium* divides inside the blood cells; the blood cells burst, releasing daughter plasmodia; fever starts as the plasmodia reproduce and increase in numbers.

Comment The increase in temperature for most illnesses starts as micro-organisms reproduce inside the host's body. This takes a few days, after which the symptoms start to show.

3 Stopping the spread of disease (page 53)

Q1 a Two from: the common cold virus mutates and changes the structure of its protein coat; the antibodies developed by the body do not recognise the new structure; so they are unable to destroy the virus.

Comment Try to extend your answer. The answer 'colds are caused by a virus' is only half the answer. Check you have written as many pieces of information as there are marks available.

b Four from: a few bacteria mutate and have a natural resistance to the antibiotic; the antibiotic kills all the other bacteria; the mutant bacteria survive; the mutant bacteria reproduce rapidly, due to little competition; a resistant population of bacteria increases in number.

Comment The important fact to remember is the bacteria are successful due to a genetic mutation. Most mutations are harmful to the organisms, but in some cases, when it is helpful, it allows the organism to survive more successfully.

c Make sure patients complete their course of antibiotics and only use antibiotics when really necessary.

Comment A common mistake is to write about preventing the spread of these bacteria by improving hygiene.

Q2 a Antiseptics are chemicals that are used to destroy microbes. Antibiotics are chemicals that can be taken internally to destroy bacteria.

Comment The important fact is fact is that antibiotics only affect bacteria.

b *Penicillium* is a type of fungus. Penicillin is the most common group of antibiotics that is produced from the fungus *Penicillium*.

Comment There are many forms of penicillin.

c Fleming, Florey and Chain.

Comment Most people can remember Fleming; however, it is important to also mention the work of Florey and Chain.

Q3 a Salting and using sugar.

Comment Salt and sugar change the concentration of water molecules in the solution.

b Water molecules move out of the bacteria into the surrounding solution. Bacteria are destroyed due to lack of water.

Comment When writing about osmosis it is important to write about the movement of water molecules. If you just refer to water your answer may become confusing.

4 Drugs and health (page 56)

Q1 Caffeine is a stimulant. Stimulants affect the synapses of neurones, causing nerve impulses to pass swiftly through the synapse. Alcohol is a depressant. Alcohol can act on certain synapses in the brain. This reduces the activity of certain areas of the brain. Alcohol can depress the areas responsible for anxiety. Larger amounts of alcohol can affect the areas of balance and conscious thought.

Comment It is important to write about the effects on the nervous system. Do not write vague answers about feelings or emotions.

Q2 Alcohol affects synapses in the brain causing your reaction time to be delayed. This could cause an accident when driving a car.

Comment Try to extend your answer. An answer such as 'alcohol affects your reaction time' will not gain full marks.

UNIT 6: PLANTS

1 Photosynthesis (page 58)

Q1 Carbon dioxide from the air enters the leaves through the stomata. Water moves from the soil into the roots and up the stem to the leaves.

Comment As plants cannot move around, their raw materials have to be readily available. Note that sunlight is not a raw material. It is the energy source that drives the chemical reaction.

Q2 Starch is not soluble.

Comment Starch is a useful storage material but needs to be broken down into smaller molecules to be able to dissolve.

Q3 Being broad allows as much light as possible to be absorbed. Being thin allows carbon dioxide to quickly reach the photosynthesising cells.

Comment There are exceptions to this principle. See page 65 about cactus spines and pine needles.

2 Transport in plants (page 62)

Q1 Xylem is made of dead cells. It carries water and minerals from the roots, up the stem, to the leaves. Phloem is made of living cells. It carries dissolved food substances from the leaves to other parts of the plant.

Comment Make sure you can also identify the xylem and phloem in diagrams of sections of roots, stems and leaves.

Q2 On a sunny day the guard cells will fully open the stomata, allowing water vapour to pass out more easily. When it is hot, water evaporates more quickly into the air spaces inside the leaf, which causes the water vapour to diffuse more quickly out of the leaf.

Comment Don't forget that if the plant is losing more water than it is taking in it will start to wilt. In this case the stomata will start to close, so reducing transpiration.

Q3 In a turgid cell the cytoplasm presses hard against the cell wall because the cell contains as much water as it can. This gives the cell rigidity. In a plasmolysed cell so much water has been lost that the cytoplasm shrinks and comes away from the cell wall. A plasmolysed cell is flexible and can't keep a fixed shape.

Comment In a healthy plant none of the cells should be plasmolysed. Remember that when cells become plasmolysed, although the cytoplasm shrinks and the membrane with it, the cell wall does not.

3 Minerals (page 66)

Q1 a Nitrogen and sulphur.

Comment Plants do not take in and use the pure elements. For example, there is a lot of nitrogen in the air (79%), but in its pure form it is very unreactive and the plants cannot use it.

b Protein is used to make new cells. For example, cell membranes contain protein.

Comment This applies to animals as well as to plants.

Q2 Magnesium is found in chlorophyll, the green pigment that gives plants their colour. Without it they cannot be green.

Comment Chlorophyll is actually chemically similar to haemoglobin in the blood, but contains magnesium instead of iron.

Q3 Minerals are absorbed by active transport. This means that energy is required, which is provided by the mitochondria.

Comment Remember that although water and minerals both enter through the roots, water enters by osmosis, which is a 'passive' process – it does not require energy.

4 Plant hormones (page 67)

Q1 Plant hormones are not made in glands. Nor are they carried in the blood.

Comment It is because of these differences that many people refer to them as growth regulators instead. Look back at Unit 4 if you need to revise about animal hormones.

Q2 a, c and d.

Comment Plant hormones affect growth and development. Pollination is simply the transfer of pollen between flowers.

Q3 Auxin is no longer concentrated more on one side than the other, so one side does not grow more than the other.

Comment *Remember that growth in tropisms is by cell elongation, not the formation of new cells by cell division.*

5 Plants and disease (page 70)

Q1 a The potato plant dies. The fungal spores attack the plant through the stomata of the leaf. The fungus spreads rapidly through the plant, destroying cells and plant tissue.

Comment *Try to write more than 'the potato plant dies'. Vague answers will not gain full marks.*

    b Crop rotation, changing sowing times, grafting, chemical treatment and selective breeding.

Comment *These methods are used to control most plant diseases, not just potato blight.*

    c Sketch similar to that found on page 70.

Comment *The diagram of the leaf structure would usually be provided for you to add your sketch of the fungus infection.*

Q2 Crop rotation, which involves sowing different crops in a field each year. Different plants are not affected by the same spores or bacteria that are left in the soil. By changing sowing times, plants are grown at times of the year when disease is not active. Grafting prevents the disease entering the plant through the roots. Chemical treatments can be used, for example spraying fungicides onto crops and soil to kill microbes. Selective breeding can be used to create disease-resistant varieties, but this takes a long time.

Comment *This type of question will be worth several marks. Check the number of marks available and then plan your answer. Write the same number of key words as marks available in the margin, to help you structure your answer.*

Q3 **One reason for** genetically modified crops: more crops will be harvested, which means more food at cheaper prices. **One reason against** genetically modified crops: people are concerned that the genetically modified plants will spread and become difficult to control.

Comment *It is important to include scientific reasons rather than emotional thoughts. During your revision, practise answering this style of question. Check your answer by underlining the scientific reasons.*

UNIT 7: ECOLOGY AND THE ENVIRONMENT

1 The study of ecology (page 71)

Q1 This is because the taller trees will be taking most of the light and with their roots taking most of the water and minerals from the soil.

Comment *This is an example of **competition**, in which the trees are out-competing the smaller plants for these important resources.*

Q2 If the numbers of rabbits went down, foxes would just concentrate their diet on other things. For example, many foxes live in towns, scavenging rubbish. If foxes died out, other things, including humans, would still kill rabbits.

Comment *There are very few other large animals in northern Canada, which is why the example of lynxes and snowshoe hares is unusual in showing so clearly the way that predators and prey affect each other.*

Q3 a The greenfly numbers would go down.

    b There would have to be enough food to feed them and they would have already eaten most of their prey.

Comment *Numbers of most species alter a little each year, but they are usually kept more or less constant because of factors like food supply and predation.*

2 Sampling techniques (page 74)

Q1 **a** Tullgren funnel.
b Pitfall trap.
c Beating tray and then a pooter.
d Pooter.

Comment Check with your teacher whether your science exam will ask you questions about these.

Q2 The average number of dandelions in 1 quadrat (1 m^2) was $25 \div 10 = 2.5$. The field is 200 m^2 so the total number of dandelions is $2.5 \times 200 = 500$.

Comment You might have to do some maths so don't forget your calculator in your exam.

Q3 If the pupil in the previous question had only placed her quadrat where there were dandelions and had deliberately missed out those areas without any she would have calculated a much higher estimate.

Comment It is for the same reason, to get a more reliable average and total estimate, that you also need to take a reasonably large number of measurements before you work out your average.

3 Classification (page 76)

Q1 **One** from: plants do not have flowers throughout a year; flowers can be different colours in different soils.

Comment Sometimes the answers may seem obvious. The examiner will never set 'trick questions'. Easy questions are often used at the start of a series of structured questions.

Q2 An artificial system of classification groups organisms according to a single characteristic. A natural system of classification groups organisms according to their most common characteristics.

Comment A common error is to write about one of the two points. When asked to comment about two things you need to write about both of them, not just one.

Q3 **a** (i) A species is a group of similar organisms that can breed with each other to produce fertile offspring.

Comment This is a frequently asked question and the definition should be learnt well in advance of the exam.

(ii) The binomial system gives each living organism two names, a genus name and a species name.

Comment The helpful hint here is that bi means two.

b **Two** from: each name is unique to an organism; it is a standard international system; no matter where you are in the world you can identify an organism.

Comment It is important to think about science world-wide, not just for your country. There are many scientists all over the world using the binomial system.

4 Relationships between organisms in an ecosystem (page 78)

Q1 **a** ladybirds
aphids
rose bushes

b ladybirds
aphids
rose bushes

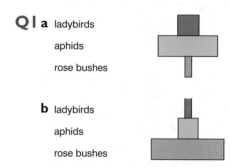

Comment In a food chain where the producers are much larger than their consumers the pyramid of numbers is inverted. However, the pyramid of biomass is always a normal pyramid shape.

Q2 It is more energy efficient to eat plant crops rather than to raise animals. This is because we are eating at a lower trophic level. At each transfer between trophic levels energy is transferred away from a food chain.

Comment **Ecologically**, therefore, it is much more efficient to be vegetarian.

Q3 Energy is lost at each stage of a food chain, so there is a limit as to how long a food chain can be.

Comment This is why pyramids of biomass are always a pyramid shape.

5 Natural recycling (page 82)

Q1 a Photosynthesis.

Comment Make sure you remember the details of photosynthesis from Unit 6.

b Respiration and combustion.

Comment Until the last few hundred years combustion played quite a small part in this. It is now playing a bigger role and this is the reason for the build up of carbon dioxide in the atmosphere.

Q2 This is the only way they can get the nitrogen they need.

Comment The nitrogen is made available by digesting the protein in the insects' bodies.

Q3 Nitrifying bacteria produce nitrates from decayed remains. Denitrifying bacteria convert nitrates and other nitrogen compounds into nitrogen gas.

Comment Denitrifying bacteria therefore have an opposite effect to nitrogen-fixing bacteria.

UNIT 8: USING ORGANISMS AND THE ENVIRONMENT
1 Using microbes (page 85)

Q1 a To kill unwanted bacteria.

Comment This question could be answered even if you have not studied about making yoghurt. The questions refer to your knowledge of bacteria and the stimulus material in the question.

b The bacteria will reproduce quickly.

Comment The usefulness of bacteria is that they can reproduce very quickly by asexual reproduction.

c At 80° the bacteria *Lactobacillus* will be destroyed so lactic acid will not be released.

Comment It does not matter that you have not heard of this bacterium before. It is your understanding of micro-organisms and their ability to use the stimulus material that is being tested.

Q2 a (i) At 0° the enzymes work very slowly, so production of penicillin would be very slow.

Comment Enzymes control all biological processes and they are affected by temperature. When a question refers to the effect of temperature it is useful to consider the involvement of enzymes as a possible answer.

(ii) At 65° the enzymes would have become denatured so the production of penicillin would stop.

Comment Never say the enzymes have been killed, they were never alive in the first place. Denatured is a useful word: try to remember it.

b The fermenter and all other equipment and substances need to be sterile. This will ensure no unwanted bacteria enter the fermenter. Oxygen is needed for the process, so sterile air is pumped into the fermenter. Stirrers then move the oxygen through the mixture.

Comment Keeping everything sterile is very important. The conditions are perfect for the growth of the majority of micro-organisms, not just the ones being produced.

Q3 a The bacteria will reproduce quickly.

Comment The first part of this question could be answered even if you have not studied about industrial fermenters. The questions refer to your knowledge of bacteria.

b Two from: bacteria divide asexually; offspring will be identical to their parents; an exact copy of genetic information is passed on each time.

Comment The usefulness of bacteria is that they can reproduce very quickly by asexual reproduction. This allows identical offspring to be produced rapidly.

c Two from: penicillin; insulin; alcohol for fuel; most antibiotics.

Comment Insulin and penicillin are the more specific answers. If you cannot remember these, then a general reference to antibiotics should gain marks.

2 Farming methods (page 88)

Q1 Space, diet and protection from predators and disease.

Comment The factors needed for fast growth will be similar for all organisms. Do not be put off if you have not studied about fish farms.

Q2 Carbon dioxide and light.

Comment This is an example of adapting your knowledge. You may not have been taught this fact; however, your knowledge of photosynthesis will allow you to answer the question.

Q3 Advantages include it could be cheaper, the farmer can be more selective, there can be less seasonal variation. Disadvantages include issues of cruelty, a possible reduction in taste or quality.

Comment Try to provide scientific reasons for your answers rather than emotional responses such as cruelty to animals.

Q4 Farmers using intensive farming methods use herbicides to control weeds. Farmers using organic methods do not use chemicals, and they need to remove weeds by hand or by mechanical methods.

Comment Removing weeds by hand or mechanical methods is very time consuming. This extra labour cost will need to be added to the price of the produce.

Q5 Organic farming methods require more people, more land per animal is needed and fewer crops are grown in smaller fields.

Comment This is an important fact to remember about organic farming.

Q6 a Biological control is the use of a living organism to control a pest population.

Comment It is important to learn the definitions of key words. This is a popular question relating to farming methods.

b A suitable method can be described taken from those listed on page 90.

Comment Try to remember two different methods. In your exam one method might be described and then you will be asked to suggest another method. This would be unfortunate if you had only revised the method used in the stem of the question.

3 Human influences on the environment (page 91)

Q1 They do not easily break down and so remain in the bodies of organisms that have taken them in.

Comment This is the reason they accumulate in animals and are passed along food chains.

Q2 The greenhouse effect describes how some of the gases in the atmosphere restrict heat escaping from the Earth. Global warming is an increase in the Earth's temperature that may well be caused by an *increased* greenhouse effect.

Comment Remember that it is the **increased** greenhouse effect that people are concerned about. A greenhouse effect is not only normal but essential for life on Earth.

Q3 This would increase their costs, which would have to be passed on to their customers, who might not want to pay more.

Comment An added problem is that, unlike pollution into rivers, the problems caused by acid rain take place a long way from the source of the pollution.

Q4 a A natural ecosystem is a naturally established habitat containing a community that is native to that habitat. An artificial ecosystem, such as a zoo, is usually made by humans.

Comment The question asks about two ecosystems. Check that your answer refers to both.

b One from: when a habitat has been destroyed; when animals become endangered.

Comment You may not have been taught this fact. This is an example of applying your knowledge.

Q5 a The species diversity index is a useful way to summarise data. It is an indication of the diversity or richness of a particular habitat.

Comment References to species diversity index have become more widely used in recent years but may not be included in some older reference books.

b (i) One from: desert; high up on a mountain; very deep water; the Antarctic.

Comment This could be anywhere where it is difficult for organisms to survive.

(ii) One from: tropical rainforests; woodland; freshwater ponds (there are many possible correct answers to this type of question).

Comment This part of the question is easier than part (i): you should be able to think of many habitats where plenty of animals and plants can be found.

Q6 a Laws need to be passed to stop people hunting animals or destroying habitats where these actions may endanger a species.

Comment Sometimes the answers may seem very obvious. Do not worry, and remember the examiner will not set 'trick questions'. Some questions will be easy for you to attempt.

b For some people in developing countries hunting is one of very few ways that people have of feeding themselves or earning money.

Comment This fact is rarely considered. It would be useful to think about these issues now and practise writing your answer.

UNIT 9: GENETICS AND EVOLUTION
1 Variation (page 101)

Q1 a A, C and E.

b B and D.

Comment If the features fall into distinct groups they are discontinuous. If they do not, they are continuous.

Q2 a Eye colour, height, shape of face (there are other examples). These are genetic features and as they are identical twins they will have the same genes.

b Skin colour, accent, how muscular they are (there are other examples). These features are affected by the environment. They are doing different jobs in different climates, so their environments are different.

Comment Some features like eye colour are purely genetically controlled and so will be identical. Some features like accent are purely environmentally controlled and so would probably differ. Many features, like skin colour, are controlled both by genes and the environment (i.e. tanning) so they would also show differences.

Q3 Chemical base, gene, chromosome, cell nucleus, cell.

Comment Look back at the diagram on page 102 if you're not sure.

2 Genes and proteins (page 106)

Q1 **Four** from: one gene is the code for one protein; an RNA template is made of the gene; RNA moves out of the nucleus into the cytoplasm; three bases must be used to code from one amino acid; a chain of amino acids is formed to produce the new protein.

Comment This type of question will be worth several marks. Check the number of marks available and then plan your answer. Write the same number of key words as marks available in the margin, to help you structure your answer.

Q2 DNA is divided into different sections called genes. One gene is the code for one protein. Each enzyme is a different protein.

Comment It is important to remember that enzymes are proteins.

Q3 All living things use the same four base codes for genes. This means the language of DNA is universal to every living thing.

Comment It is often difficult to answer this style of question. You will find it easier if you use the key words, such as four base codes, genes and DNA.

Q4 The four bases are put together in different combinations

Comment Students often leave this blank in exams. They have difficulty in wording their answers. Try writing out your answer during your revision.

3 Inheritance (page 107)

Q1 Let R be the dominant allele for tongue rolling and r the recessive allele for not being able to roll your tongue.

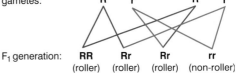

parents: Rr   Rr

gametes: R   r   R   r

F$_1$ generation: **RR**   **Rr**   **Rr**   **rr**
(roller) (roller) (roller) (non-roller)

Comment If not being able to roll your tongue is a recessive condition then the 'non-roller' child must be rr. This means each parent must have at least one r allele. As the parents can roll their tongues they must therefore both be Rr.

Q2 Cross the spotted leopard with a known ss leopard. If there are any black cubs then they must be ss and have received one of the s alleles from the spotted parent, which must therefore be Ss. If there are no black cubs the spotted parent is likely to be an SS. The random nature of fertilisation means that an Ss leopard may still produce no ss cubs, but this is unlikely with a large litter.

Comment Crossing with a homozygous recessive is known as a test cross or back cross. It is used because the homozygous recessive is the only genotype that can be identified by the phenotype alone.

Q3 As neither has the condition both must be heterozygous carriers of the recessive allele if their child has two copies of the recessive allele. The probability of a child being homozygous recessive, i.e. having cystic fibrosis, is 1 in 4.

Comment Look back at the monohybrid cross shown on page 108. You can see that the offspring showing the recessive condition are in the ratio 1:3 to those showing the dominant condition. Another way of saying this is that there is a 1 in 4 chance of a child being homozygous recessive and showing the recessive condition. The fact that the couple have one child with the condition is not a consideration.

4 Applications (page 111)

Q1 Pick sheep with finer wool than the others. Breed these together. From the offspring pick those with the finest wool and breed these. Continue this over many generations.

Comment The principle of selective breeding is the same whatever the example.

Q2 Advantages: many roses with identical flowers can be produced relatively quickly. This would be useful if the roses on this particular plant sold well. **Disadvantages**: there is no variation, so if one rose was attacked by a plant disease then all the rest would be susceptible too.

Comment The key point about cloning is that the new plants are genetically identical. This may or may not be an advantage.

Q3 New combinations of genes can be produced quickly and reliably with genetic engineering. Genetic engineering allows genes to be taken from one species and given to another, which could not be done with selective breeding as different species cannot breed together.

Comment The second answer is probably the more important. For example, a gene for resistance to a particular disease could be taken from one species and given to another.

5 Evolution (page 115)

Q1 a Vertebrates have hard body parts, e.g. bones and teeth, which can leave fossil remains, whereas many invertebrates have only soft body parts which rarely leave fossils.

Comment This does not apply to all invertebrates. There are many remains of the shells of marine invertebrates. Even so, these leave little evidence of what the soft tissues were like.

b There are many reasons, such as: the remains of the animals did not lie undisturbed in the right environmental conditions to form fossils, fossils were formed but have been destroyed – e.g. by erosion – or the fossils have not been found yet.

Comment Although the fossil record of evolution is incomplete and may not give detailed information about how all living things evolved it is still a very important source of evidence.

Q2 Among the shorter-necked ancestors there would have been *variation* in height. A giraffe that was only slightly taller than others would have been able to reach and feed on some leaves that others would not have been able to. They would have had a slight *advantage* in the *competition* for food and so would be slightly better fed and slightly more healthy than shorter ones. They would therefore probably on average live a little longer and be able to successfully rear slightly more offspring. If they were able to pass on the longer neck to their offspring because it was a *genetic* feature, it would mean both that there would be more of their offspring than of the shorter-necked giraffes and that the overall average height of the giraffe population would have increased. If this continued for many, many generations it would lead to the giraffes we see today.

Comment In an exam, questions about natural selection will often require extended answers and may carry four or more marks. To get full marks you will need to give a detailed response. Do not be put off if the example is unfamiliar. This is to test that you understand the **process**, which will be the same in each case. Note that natural selection favours features that give an advantage but will not always result in evolutionary change. In the example of the giraffe, being taller than they are today would pose disadvantages, e.g. the difficulty of pumping blood to the brain or problems of stability, which means that they will not just continue to get taller. Some animals alive today are known as 'living fossils' as they appear similar to ancient fossil remains. If an organism's environment has not changed significantly and it is well adapted to it, then natural selection would work to keep the animal's features the same.

UNIT 10: EXAM PRACTICE
Foundation Tier (page 124)

Q1 The following words in this order:
decreases
decreases
increases
stays the same

Examiner's comments
For this style of question you can use the words in the box once, more than once or not at all.

Q2 a 94.8

Examiner's comments
You are not expected to remember this amount. You would be expected to calculate the amount from the data provided. Remember when it states percentage the numbers should add up to 100.

b (i) To cool the body or to maintain a constant body temperature.

Examiner's comments
A vague answer such as, 'to let heat out' would not gain marks.

b (ii) Water and ions.

Examiner's comments
Both are needed for one mark. Water is the easy part, you then need to decided between sugar and ions. Sugar is needed by the body cells for respiration.

b (iii) Water

Examiner's comments
A common error is to write carbon dioxide. Of course carbon dioxide is lost from the body during breathing but it is not the correct response for this question. Make sure you read the questions carefully.

c Any two from: sugar is used in respiration; to provide energy; energy is needed for movement such as running.

Examiner's comments
This is worth two marks. Check your answer to make sure you have provided two pieces of information.

Foundation/Higher Tier Overlap (pages 130–131)

Q1 a **Two** from the following: they do not produce insulin for themselves; their pancreas is not working correctly; they need to control their blood glucose level, to stop it rising too high; the insulin would be digested if taken orally.

Examiner's comments
There are two marks for this question. You must check you have included two pieces of information in your answer.

b **One** from the following: people eat meals and take in glucose several times a day, not just once a day; one large injection may reduce glucose levels too much; after several hours the body would digest the insulin.

Examiner's comments
You may not have been taught this fact. Some exam questions will expect you to apply your knowledge and think about the possible answer. Your understanding of the subject should be sufficient for you to apply your knowledge and work out the correct answer.

c **Two** from: insulin is a protein; it will be digested by pepsin (or protease); it could be denatured by stomach acid.

Examiner's comments
Do not use the words 'destroyed' or 'killed' when referring to enzyme activity. Try to use the correct names for the enzymes. A vague answer such as 'it will be digested by enzymes' will not gain credit on a Higher Tier exam paper.

d Two from: the liver is the target organ; the insulin will increase the permeability of cells to glucose; the insulin will cause the uptake of glucose from the blood.

Higher Tier (pages 134–135))

Examiner's comments
You could also write about the conversion of glucose into glycogen and glycogen being stored in the liver.

Q1 a (i) Arrows drawn on diagram from left to right

Examiner's comments
When you are asked to draw on a diagram, make sure your markings are clear and easy for the examiner to see.

Q2 a Bacteria was passed on from dirty clothes, hands or equipment.

Examiner's comments
Do not use the word 'germs', this word is only expected from infants. 'Bacteria' and 'pathogens' are more suitable words to use.

   (ii) The neurone has a myelin (or fatty) sheath. The neurone is very long and it has many projections and many nodes.

Examiner's comments
It is very common to have a question asking about how parts of the body are adapted for their function.

b After Lister's instructions, fewer people died. The chart shows there is a reduction in the number of patients dying from infections after operations.

Examiner's comments
The first sentence is only a vague comment and will only gain one of the two marks available. Always try to extend your answer.

b (i) The hormones travel in blood.
   (ii) Hormones have a slower response; hormones have a longer lasting effect.

Examiner's comments
Another correct answer would be that hormones have a more widespread response.

c (i) Use of a kidney dialysis machine.

Examiner's comments
The question refers to kidney transplants and asks for one **other** way. Always read the questions carefully and look for key words.

c (i) **Three** from the following: at the start the growth is the same on both sides; the growth on the shaded side speeds up; the growth on the side with light shining on it slows down (or stops); the side with light shining on it speeds up again after a while. One extra mark is for a clear, ordered answer.

   (ii) A donor kidney is specially chosen, to ensure a match of a similar tissue type. The recipient's bone marrow is treated with radiation, to stop the production of white blood cells. The recipient is treated with drugs, to suppress the immune system. The recipient is kept in sterile conditions, to avoid exposure to pathogens.

Examiner's comments
You need to plan your answer to questions carrying more than two marks. Write down three key words in the margin that you need to use in your answer. As you write your answer, include each key word, crossing it out when you have written about it. Questions carrying the symbol for marks available for the quality of the written answer must always be checked. For this question you must check that your answer comments on the growth of both sides of the stem.

Examiner's comments
The four marks is an indication that there is one mark for each precaution; Try not to write too much for one precaution; this may cause you to be too brief on the others.

(ii) **Three** from the following: auxins are involved in the process; auxins are produced in the tip of the shoot; the additional light causes more auxin to be sent down the shaded side; the extra auxin causes faster growth to occur on the shaded side.

Examiner's comments
The word 'explain' in the question means you need to write about the scientific reasons why this process happened. Check your answer to make sure you have not written a description by mistake.

(iii) This response allows plants to grow towards the sunlight (or light). The plant will receive more light for photosynthesis.

Examiner's comments
'Receiving more light for photosynthesis will ensure more growth' would also be a correct answer.

Q2 a Glucose is an end-product of digestion.

Examiner's comments
You could also have explained that glucose in the small intestine is absorbed into the blood.

b A hormone is a chemical messenger.

Examiner's comments
This is a frequently asked question. Try to learn the definition.

c (i) Glycogen

Examiner's comments
You may not have been sure about the name of the polymer, but you should understand that glucose changes into glycogen.

(ii) Insulin is transported in the blood steam.

Examiner's comments
Most substances are transported around the body in the blood stream. If you are not sure of the answer, make sure you attempt the question. A well thought out guess may gain credit.

d (i) The instructions travel as electrical impulses in neurones.

Examiner's comments
If you are not sure about the answer, think about the key words in the question. The brain is part of the central nervous system so you should be thinking about the nervous system.

(ii) $6H_2O$

Examiner's comments
If you are attempting the Higher Tier exam, you need to learn the symbolic equations for respiration and photosynthesis.

(iii) **Two** from: aerobic respiration transfers more energy; aerobic respiration is more efficient than anaerobic respiration; the waste products of aerobic respiration are less toxic than those of anaerobic respiration; anaerobic respiration does not produce lactic acid. There is usually an adequate supply of oxygen so there is no need for anaerobic respiration.

Examiner's comments
This is a very common question.

organic farming 90
osmosis 5–6, 58, 63
ovaries 38
oxygen 8, 9, 11, 17, 58, 82
oxygen debt 9
oxyhaemoglobin 10, 11
ozone layer 96

pancreas 23, 38
parasitism 73
Pasteur, Louis 47
pathogens 47, 49, 51, 55
Pavlov, Ivan 21, 35
penicillin 54–5, 86
peristalsis 22, 23
pesticides 92
phage technology 55
phagocytes 46
phloem 60, 62
phosphates 66
photosynthesis 2, 58–61, 78, 82
    optimum conditions 61, 89
    and respiration 60
pituitary gland 37
plankton 27
plant biology 58–70
plant cells 1, 2
plasma 10
plastics 84
platelets 10
pollution 91–7
population 71
    growth 72–3, 91
potassium 66
predators 72, 78, 92
prey 72, 78
producers 78
progesterone 38, 39
proteins 8, 19, 82
    in DNA 101, 102, 106
    meat-free substitute 86
protozoa 49
Punnett square 108
pyramids (ecology) 79, 81

quadrat 75

rainforests 97
Ray, John 76
receptors 31
recombinant DNA 106
recycling 82–4, 91, 97
red blood cells 5, 10, 11, 23
reflexes 34–5
reproduction
    asexual 103
    sexual 104, 116–17

reptiles 30
respiration 2, 8–9, 81, 82
    in plants 58, 60
respiratory tract 46
RNA (ribonucleic acid) 106
roots 63
rubbish tips 97

Salmonella 53
sampling techniques 74–5
scurvy 19, 20
selective breeding 111–12, 113
sewage 52, 84, 97
sex determination 110
single-cell protein (SCP) 85, 86
skeleton 28
skin 46, 54
smoking 17, 56–7
sodium hydrogencarbonate 22, 23
solvents 57
species 71, 77
    affinities between 115
    endangered 98–9
    evolution into 117
species diversity index 97
spinal cord 32, 34
starch 58, 59
stomach 23
stomata 60, 64, 65
sucrose 58, 59
sulphur dioxide 95
sunlight 58
sustainable development 100
synapses 35, 56
system 2

tapeworm 50
taxonomy 76–7
temperature
    of Earth 94
    homeostasis 41–2
    in photosynthesis 61
tendons 28
testes 38
testosterone 38
thorax 15, 16
thyroid gland 37
tissue 2
tissue culture 112
tobacco 17, 56–7
toxins 46, 51
Toxocara 50
trachea 15
transect 75
translocation 62
transpiration 63–4, 66

transport 4
    by blood 10–14
    and cells 4–7
    in plants 62–5
trophic levels 78, 79
tropisms 67–9
turgid cells 5, 64

ultrafiltration 43
ultraviolet (UV) radiation 96
urine 43

vaccination 47, 51
vacuole 2
variation 101, 111, 113
vegetarianism 81
veins 11, 12, 13
ventilation 15, 45
vertebrates 28
villi 23, 24
viruses 46, 49, 50
vitamins 19, 23

Wallace, Alfred Russell 115
water 52, 58
    balance 42–5
        monitored 44–5
    vapour 17
Watson, James 102
weedkillers 67
white blood cells 10, 46

xylem 62, 63

yeast 50, 85

zygote 107

William Collins' dream of knowledge for all began with the publication of his first book in 1819. A self-educated mill worker, he not only enriched millions of lives, but also founded a flourishing publishing house. Today, staying true to this spirit, Collins books are packed with inspiration, innovation and practical expertise. They place you at the centre of a world of possibility and give you exactly what you need to explore it.

Collins. Do more.

Published by Collins
An imprint of HarperCollins*Publishers*
77–85 Fulham Palace Road
London W6 8JB

Browse the complete Collins catalogue at
www.collinseducation.com

© HarperCollins*Publishers* Limited 2005

First published 2005

10 9 8 7 6 5 4 3 2 1

ISBN 0 00 719058 1

Jackie Clegg and Mike Smith assert their moral rights to be identified as the authors of this work.

British Library Cataloguing in Publication Data
A catalogue record for this publication is available from the British Library.

Acknowledgements
The Authors and Publisher are grateful to the following for permission to reproduce copyright material:
AQA: questions 4, 8, 9
Edexcel: questions 2, 3
OCR: questions 1, 5, 6, 7, 10, 11, 12

Photographs
Andrew Lambert: pp. 103, 112T; Claude Nuridsany & Marie Perennou/Science Photo Library: p. 92; D Phillips/Science Photo Library: p. 54T; Erika Craddock/Science Photo Library: p. 51; Getty Images: p. 90; Hank Morgan/Science Photo Library: p. 89; Holt Studios International: pp. 73, 93; John Durham/Science Photo Library: p. 54B; Lee D Simon/ science Photo Library: p. 50; Martyn F Chillmaid?Science Photo Library: p. 67; Natural History Photographic Agency: pp. 77, 111, 117, 118R; Nature Picture Library: p. 71; Peter Menzel/Science Photo Library: p. 88; Science Photo Library: pp. 11, 97, 102, 104, 112B, 118; Taxi/Getty Images: p. 116; Wally McNamee/Corbis: p. 9

Illustrations
Roger Bastow, Harvey Collins, Richard Deverell, Jerry Fowler, Gecko Ltd, Ian Law, Mike Parsons, Dave Poole, Chris Rothero and Tony Warne

Every effort has been made to contact the holders of copyright material, but if any have been inadvertently overlooked, the Publishers will be pleased to make the necessary arrangements at the first opportunity.

Edited by Eva Fairnell
Series and book design by Sally Boothroyd
Index compiled by Ann Lloyd Griffiths
Production by Katie Butler
Printed and bound by Printing Express, Hong Kong

You might also like to visit
www.harpercollins.co.uk
The book lover's website